Lucia Dettori

Delta
The Law of Dimensions

Between Science and Spirituality
To Generate
Harmony Knowledge Balance Beauty Love

Delta The Law of Dimensions

ISBN-13: 978-1530957132

ISBN-10: 1530957133

DELTA   The Law of Dimensions

PREMISE

Everything becomes possible for you if you learn to live in light and of light. Everything is light, you are beings of light and your life is wonderful.

Light is all around you and within you. Happiness and joy, love and beauty, are for you and within you.

The path towards finding them is easy; you just have to want to walk it. Begin! Take the first steps without wondering which way is better, because all roads merge into one and the direction is always the same. The ways of being light in light are infinite and are before you: walk them, because the time has come.

Change is beauty and it is happening now, before your very eyes; watch it carefully, without fear. It opens the doors to light; pass through them and your joy will be immense. Everything will be harmony, pure and infinite, emotion and life.

Each and every one of you is a unique and unrepeatable being, different from everyone else and full of potential. Once you have remembered this, the path has already begun.

The path is the beginning of the light....

Lucia Dettori

INTRODUCTION

All those who find themselves before these pages at this point in time are people who - whether they are aware of it or not - are questioning themselves about the sense of their life.

Some because they are suffering in body or in spirit, others because they are close to people who suffer, others again simply because they ardently want to learn something different that goes beyond what they have encountered until now.

I too have asked myself these questions and have done so with great insistence over the past ten years.

They have been years of research, of study and of learning on many levels.

I have found the answers I was looking for and have identified a method by which everyone - if they want to -can arrive at their own answers.

One of the things that you learn when you reach a certain point in knowledge is that *"whoever finds must tell"* as many people as possible; they must *Pass the Information*. That is what led me to write this book, so that I could share everything that I found with you and show you the path that can be travelled by means of this rediscovered instrument.

There are answers to all your questions and very good ways to change your life, giving it the sense it has always had, but that has escaped you for a long time now.

The many facets of reality and the various lines of quantum possibilities onto which they decide to shift lead human beings to experience the same world in completely different ways.

Every way, every path and every technique that has been discovered and disclosed in recent years is equally effective; in fact, each of them suits different people. In virtue of the many quantum possibilities available in reality and despite their being joined in their search for personal evolution, people need to choose different ways, according to their own desires, expectations, speed of change and degree of awareness.

Let us accept, therefore, the movement and variety of ideas, theories and solutions, because everything is heading towards a single focal point that everyone will reach, in their own time and according to their own particular characteristics.

Bring on the energy of change, bring on the Light!

These pages will present a different kind of method, because it is addressed to a different kind of person.

There is an infallible way of knowing whether this is the "method" you are looking for: open the book at random and read a few lines. If your heart feels an immediate connection, buy the book; otherwise simply put it back in its place, it will be for someone else.

CHAPTER I

THE PATH

It seems useful, for purposes of a better understanding of all the issues contained herein, to dedicate this first chapter to illustrating the fundamental stages that led to such different, yet such ancient, knowledge. Knowledge that is entirely new, yet permeated with as many traits and ancient archetypes as human existence itself.

The new, even when it is absolutely new, always and anyway originates from the old, particularly when you want to move as far away as possible from the known.

It is wise therefore to avoid excluding the old, because it is so rooted in human memory that it will inexorably return to place the emphasis on the consequentiality of everything.

Going against what has already been means going against a part of yourself. Beginning from what has been, to process the new, is evolving. Evolution of this kind is possible thanks to anyone who contributed towards it, and those who came after benefit from those who came before. This applies even when you find that sometimes the knowledge of those who came before was greater than the knowledge that followed. Knowledge exists even in those who have never had conscious access to it. Latent, dormant and distant, yet concrete and easily attainable, It exists within them, because it was passed down by those who came before.

1. The other reality.

I have never been able to say exactly when my research path began, but whenever I think about it, I realise that everything I have ever done in my life has led me towards this.

Since I was a child, I was absolutely certain that my day to day life was not the Real Life, as I called it then. I often happened to live such a convincing degree of materiality in that other reality as to be truer than life. I no longer know how explain it properly, but at that time it was perfectly clear to me; today, in adult words, I can say I perceived a parallel world. I lived in that world and also in this, understanding the differences between them. I do not know if it was the typical imaginary world of children and, to tell the truth, to this day I do not know how much is imaginary or real in the world of children; they do not pose the question, they just live.

One thing is for sure: that world has never ended for me; it continues.

As an adolescent, I had an abrupt "awakening". My friends accused me of being too different, and so I adapted. I created two separate and distinct worlds for myself: an outer one, made of day-to-day things, of girlish interests, of famous singers, of favourite teams...and an inner one, made of books, places, emotions, feelings, lives, landscapes and unknown yet, to me, strangely familiar times.

Since then, my favourite books have always been historic novels; it fascinates me to learn how people lived during the course of centuries and civilisations. A kind of literature that I discovered by chance (I now know that chance does not exist) when, at the age of twelve, after having read all the children's books there were in the house, I came across Tolstoy's War and Peace. It was love at first sight for me, that book opened my eyes onto what I was searching for: history narrated from a different perspective.

I learnt, as I imagined, that everything in reality can have different interpretations; I have never forgotten that lesson. I have continued to see everything from various perspectives. That novel was what I needed; it represented the "scientific" consecration that allowed me to continue to cultivate the different objectivity that I had spontaneously sensed.

After that, the biographies of great historic personae have always helped me to understand the other side of their reality by myself, a reality that differs from that told in history books. I sensed the magic of their "greatness", and the thirst but also the thirst for life -often for destruction- that drove them.

Unfailingly I would be faced with a truth that continuously showed me: there is never just one truth. By reading attentively, I discovered, for example, that Alexander the Great, who was celebrated in the West as the Great Conqueror and worthy representative of that human passion that burns life in the pursuit of a dream was, and still is, to the peoples of Central Asia, Iskander the Accursed, who brought death and destruction and drenched ancient lands of knowledge and memory in blood.

Two truths, one man.

I proceeded with my search for reality, keeping my two "lives" separate; I attended my traditional schooling during my day-to-day life, but set aside time for my other life, in which the books I read gradually turned into actual parallel studies that hinged primarily on certain fronts: ancient Egypt, the old European religions, everything that concerned the other truth about Jesus' life, and the ancient religions of Central Asia.

Disparate interests, it seemed to me at the time; a single, great path, I know now.

In my little town, my high school years passed peacefully, reading and studying. There were no particular upheavals during those years, nor during the following ones when I moved to Florence to complete my studies. Both my lives continued to flow parallel and alive: I felt I was at the centre of the world and I learnt an enormous amount, especially from the people I met. Once I graduated I decided to go back to Sardinia to work as a self-employed architect. I felt that despite my love for far-away lands, that was where I wanted to do something, even if I did not know exactly what.

Being self-employed has always been of utmost importance to me. By nature I am contrary to anything that gives me the idea of a lack of freedom. This professional choice was highly gratifying to me from the creative point of view and, at the same time, it left me time to live my other reality made of books, travel, constant research...

I was able to move deftly between the two realities and had actually progressed to formulating a theory of my own, then in its embryonic stage. Basing myself on the assumption that everything we experience in our daily lives is created by our minds (and, because I "knew" the other reality, I could affirm with certainty), by deduction I had arrived at stating that even malaise does not really exist, but that this too is a conventional creation of the mind. I was sure that once freed of such beliefs, everyone would have lived without pain.

There was and is a precise reason for which my attention focused on sickness and on the possible ways of removing it from our daily life, and that was that since before I was born my mother's life was "at risk", according to medical science. My life, therefore, is a continuous race against time. Today, thanks to the knowledge I acquired, I have made my speed a virtue, determining the direction in which my research evolved.

In the upcoming pages you will see how anything, even "dissonance", can be turned into something useful for oneself and for others.

This was my theory then, but because it had been formulated in the other reality, I kept it well away from my daily life and, especially, took care not to talk about it with just anyone. Despite this, however, I paid great attention to perceiving external signs. One day, I happened to be reading a flyer when my eye caught on a sentence that inferred there were scientific studies that confirmed a close interconnection between physical malaise and the brain. It appeared that those studies had led to the conclusion that any kind of infirmity was determined by mechanisms of the brain which, amongst the various responses to external conditioning, also included precisely that of sickness itself.

My reaction was one of immediate joy. I realised I was not crazy, and then I smiled at the thought that maybe in any case I was, the only difference lying in the fact that there were other crazy people out there just like me.

I therefore decided to learn more about this strange theory by studying the behaviour of the brain and placing it in relation to its surroundings, intended not only in terms of current environment, but also of heritage of the past.

I learnt a great deal during those years.

I confirmed the importance of observing reality from different perspectives and learnt to understand the so-called biological memories that lead human beings to live a life that is not entirely theirs. Memories passed down from one generation to another that drive one to act according to preconceived schemes to which there are pre-established -and therefore automatic- responses of which one is unaware. I learnt that sickness, behaviour, events, coincidences, are none other than a great weave of rhythms and cycles within which human beings move, and always have. It was easy for me to identify the correlation between cause and effect and, once I had seen the evident symptom, I enjoyed tracing back to the triggering cause. Everything was simple, almost mechanical, and the human brain seemed like a gear that I knew every part of and of which I was able to predict the reaction to a certain stimulus. I reached the point of understanding my fears, the cause of my brain's response mechanisms and of my behaviour; finally, I learnt how to master my life.

At the end, I felt like a new person, capable of tackling the next stretch of the path. I had learnt to "read" people simply by observing the external aspects of their life: the shape of their body, how they moved, their habits, their voice, their way of speaking, their car, house or favourite program…in other words, to read what in technical terms are known as the "manifests" of a person, and to master the techniques that help to resolve the mechanisms triggered by fear. I also learnt something else, and that was that the basic mechanism, triggered by the brain and that leads to malaise, at whatever level this may manifest, can be modified but not definitively removed. One can learn to identify the reason underlying a certain malaise and resolve it in as short a time as possible, but one cannot change the mechanism that is triggered by that particular reason, because it is structural; in other words it is part of the brain's structure itself.

I was pleased about the great understanding I had reached but, to me, the fact that the structure could not be changed, made it seem as though the brain was a machine, the metal gears of which made it rigid and for unsuited evolution....

This sounded strange to me but, believing it was part of that other life of mine -which I continued to keep separate from my daily life-, I did not mind this strangeness. At the end of the day, I was an architect who took pleasure in learning more about life and observing different things, just for my own personal growth. I was neither a psychologist, nor a psychiatrist, nor a doctor, nor anything else that had to do with such things.

Despite my looking for justification to avoid looking further into this understanding, I realised I had learnt to think differently even in my day-to-day life and felt that my two realities were drawing closer to each other at a far greater speed than I had imagined.

Today I know I had activated such an energetically strong mechanism that only my recklessness of the time could conceive of keeping even the tiniest part of what surrounded me from it. I was not aware of it, but had begun asking myself direct questions about every aspect of my life and searching for the most immediate answer, often crossing the barriers of rational thought.

I had thus become sure of one thing: in my existence I aspired to evolve in relation to every single aspect of myself. I still did not know how I would have applied all this to all the sectors of my life, but I sensed that I would have found the way. I was beginning to adopt the mindset by which perhaps these two realities of mine were not actually that separate and distinct.

Just as I began to acknowledge the need to combine my realities into one, I did something instinctively that there and then I did not fully understand: after having thanked the Universe for the opportunity, granted to me from studying these techniques, I decided I had to undertake a new path, an entirely different one. I did not know why I did so, but I sensed I needed to look for something more.

My studies proceeded in an apparently haphazard manner: manuals of human anatomy and treaties on classical physics alternated with discoveries by scientists who -in turn- called their disciplines new medicine, new genetics, new science…as though to emphasize their distance from classical science. And again: treaties on prayer and shamanic theories, bio-geology, Celtic myths and legends, archeo-astronomy…. Everything came together in my mind and my notes came to life from the pages accumulated on my desk. The extreme diversity with which the concepts were explained did not interest me, because I realised that all the knowledge I was acquiring was leading me in the sure direction of the existence of a reality that was much vaster and more complex than what the human beings of this moment in history were used to imagining.

Finally I found the lowest common denominator that linked those apparently disparate studies: all the theories, the techniques, the sciences, the meditations…all led towards a single focal point: all paths are one.

Every new discovery confirmed my conviction: all things were One. This conviction has never faltered in me, and it is from this conviction that I have gained the most benefit, because it places me in a situation of lightness, allowing me to take and to learn from everything.

At a certain point, however, I came to a stop.

After six years of study and research, I was forced to a halt because I found that all disciplines had one particular point in common, an axiom that, from my point of view, did not allow any way forward.

Every science, theory and discipline of any kind and nature seemed to find only one possible solution for the wellbeing of human beings; this solution can be summarised in the phrase *"look out"*. This meant that once the cause for a person's suffering had been identified (be it in the form of anguish, lack of harmony, anxiety, existential malaise, sadness, poverty or anything else), one could not detach oneself from it. According to the majority of theories, the cause for the malaise, which more simply can be referred to as "dissonance", cannot disappear, because it will inexorably present itself over and over again in the person's life. The only solution that all disciplines had -whether old or new - was to "look out", in other words, be aware of the dissonance as it arises and make sure it lasts as short a time as possible, intervening in various ways in order to remedy it immediately.

This is the solution I found everywhere, both in relation to "new" sciences and to ancient texts.

The ways they indicated to keep dissonance away varied slightly and, according to the discipline in question, involved meditation, prayer, the mere attentive observation of the world around us, or an actual battle as some shamanic traditions called for....

But apart from the various solutions they indicated for lowering the threshold of dissonance, the least common denominator that united these disciplines remained the fact that human beings had to deal with dissonance throughout their whole life on Earth.

Despite their being born with infinite potential, people in some way seemed to be destined -due to themselves or their fellow creatures or the environment around them, or otherwise to beliefs, convictions, traditions or memories both learnt and inherited- to have to be on the alert, constantly making sure they raise the least possible resistance and let themselves be sweetly carried along by the flow of life.

As I reflected on this, a profound sense of sadness came over me at the thought of this human condition, though I knew there was nothing I could do to change it. I had in fact acquired sufficient knowledge to understand that it you cannot in any way interfere with another person's free will. On the other hand, I was absolutely certain that one can and one must change oneself.

I cherished everything I had learnt about myself, I knew what my biological needs were, what my survival strategies were, what my sense-project and life goal were... All the understanding I had acquired about myself prevented me from thinking in a sole direction and, in the same way, prevented me from taking into consideration the sole possibility by which I was to spend my entire life paying the utmost attention, meditating or doing other similar things... From quantum mechanics, I had learnt there are infinite solutions to every event, and I sensed that this might be only one of them. I also had the strong feeling that all the methods offered for keeping conflict away were excellent, but that none of them were for me, because they aimed at achieving a solution I did not feel applied for my own personal choice of life. I imagined myself intent on concentrating all my attention on making sure I did not enter into my own personal dissonance. I saw myself focused on leading a pleasant life, in the constant search for balance and harmony, in an exhausting slalom between the poles of life itself; in that way, I would not have had the time and energy to do anything else.

I had other things in mind for myself.

I wanted wellbeing at all levels, always and

constantly, without having to look for and reconstruct it time after time. I wanted to conduct my life without paying the utmost attention, free from the continuous energetic strain of controlling the reality around me or of being afraid of losing the benefits acquired because of some distraction of mine. Wellbeing at all levels had to become a given fact in my life, acquired once and for all and always in perfect balance. This would have enabled me to move forward and dedicate myself to doing other things.

I was and still am convinced that the goal of all human beings is to evolve and that wellbeing, joy, financial prosperity, a fulfilling sentimental relationship, a gratifying occupation...are the starting points and not the arrival points of their path.

I made up my mind: if I did not find what I was looking for in other people's studies and writings, it meant that a solution of this kind was one of my own personal needs ...and that thus I had to find it on my own.

I decided to find it.

I would deliberately take the quantum leap and choose to live the possibility of freedom from any opposition; therefore harmonious balance in all the sectors of my life and at all levels.

I liked this idea a great deal, for two reasons especially: on the one hand, it allowed me to resolve what until then seemed to be a problem that only I had and that therefore nobody else seemed interested in resolving; on the other, it allowed me to apply myself to the research that was now enthralling me more and more.

I set out on the research that changed my life even further. The new path had by now assumed prime position in my daily life. For two years I ceased performing my profession -despite my loving it then as I still do now-, and immerged myself in this new research. I spent about twenty hours a day studying, and used the other four to sleep: nothing else mattered to me, just my passion.

It was in fact undiluted passion: passion that, by not answering to any known laws or rules, assumes the form it deems most fitting for itself, often reaching totalising heights as it did with me.

That it was passion, beautiful and fiery, total and pure, alive and intimate, I was certain when, once I had found the instrument I was looking for, harmony and balance began to flow through me, together with the new awareness…

2. The quantum leap.

I had therefore made up my mind: *deliberately take the quantum leap.*

At this point it is essential to clarify what I mean by that term and to do so I will digress briefly in order to illustrate one of the ways in which reality "manifests" according to quantum physics.

A quantum is the minimum defined and undividable value of a physical magnitude that can only vary by multiples of that value. It is therefore the minimum quantity of "matter" sufficient for laboratory study. According to quantum physics, all of reality - if observed in its "manifestation" in the form of particles and not of waves - consists of an infinite number of light quanta, which are referred to as photons.

Light quanta therefore create reality.

Imagine observing a sequence of these luminous dots that follow one after the other, creating extremely fine threads. Different possibilities of life exist on each of these threads which, precisely, are called quantum possibilities. Reality therefore consists of an infinite number of photon trails, which run like parallel lines, each one carrying a quantum possibility.

Quantum possibilities are also therefore infinite. It is on this principle that the theory of quantum physics is based. According to this science, in fact, there are multiple possibilities for every single event; in other words, every specific circumstance can have a different outcome. These possibilities have all already been achieved on different photon trails. This means that every possibility has already been created and is present in our world and that if you want to pass from one outcome to another, you can do so by shifting from one lane to the other or, so to speak, from one photon trail to another; this shift is known precisely as the *quantum leap.*

What I intended to do therefore was the following: shift from the photon trail on which my life had been running until then, -and on which by now I was living uncomfortably as a result of the presence and cyclic repetition of my various dissonances-, onto the photon trail on which my life was free of dissonance of any kind and of any cellular memory inherited or acquired externally, and that therefore was not mine.

I had clear in my mind the quantum possibility

trail I was searching for and on which I had chosen to live. I was in fact searching for the trail of photons on which cellular information could be changed; that trail encompasses a whole series of corollaries, not the least of which being the possibility of changing all information.

All information, with no exception.

This line of reasoning encouraged me to pursue my research, yearning to reach my goal in the shortest space of linear time. Needless to say, by nature I tend to accelerate, whenever I feel it is useful for me to do so.

Despite my great enthusiasm, however, I felt like a fly that had smacked into a window right when it seemed to glimpse its way out. Quantum science, in fact, has not yet found or, if it has, has not yet explained and disclosed how one can take the quantum leap.

Up until now, one has relied on a sort of randomness, which sometimes happens and at others doesn't. For instance, human beings have coined the term "luck" to explain the quantum leap that some manage to take at times. A multimillion win can change people's lives, even up to the smallest details of their everyday life. Not only can it change their relationship with money, but also with their job, with their personal relationships, etc. On closer inspection, however, luck is depicted like a blindfolded goddess who, fleeting and fickle, suddenly strikes to then just as suddenly flee, thus proving her total and absolute "chance".

Even a serious earthquake can radically change a person's life. Suddenly without home, possessions and sometimes even family, the person finds himself having to invent a new way of life. And because human beings always have the tendency to judge, it will be called a tragedy.

If, however, one were to look at it from the point of view of profound change, regardless of whether it was a tragedy or a stroke of luck, one could say that a quantum leap was made.

To photon trails, the concept of good or bad does not exist; they simply are and, trusting oneself to chance, one finds oneself walking on some trails rather on others, it does not matter which.

I had chosen to live differently, and I knew that the solution of *how* to existed; I just had to find it.

I started off on the right foot, because I knew what I was looking for: the way to change my life, the one on this side of the veil.

I must admit I had an extra chance of succeeding

in my discovery: my knowledge of that other reality, in which all things were revealed to me so I could understand them easily.

I also had a precious aid in all the tools I had learnt to use through my studies. And again, the endeavour enthralled me and this was another element that can be of great help in similar cases…. I thus set out.

From the very beginning I had a strong sense that the solution to what I was looking for lay in the cross-point between science, taken in the literal sense of the word, and what I simply refer to as Spirituality.

When I talk about Spirituality, I refer to that which is intangible to human beings, in that they are unable to perceive its existence through their five senses. However the intangible part - or the unknown part as one can call it, in that it does not manifest according to the usual standards - corresponds however to 90% of the Universe, and it is therefore impossible for humanity to detach from it.

According to the researchers who used the computer to simulate the creation of our universe and the so-called "Big Bang" from which it originated, 90% of the Universe disappeared shortly after the instant of the explosion.

In other words, considering a mass of matter

amounting to 100, immediately after the explosion only 10% remained. Where did the rest of it go?

By studying our reality we know that quanta vibrate at different speeds, thus giving us different consistencies of "matter" that, for simplicity's sake, I will define as being more or less compact. Therefore, by way of this definition, I can say that rocks have a low vibration, and are therefore deciphered by human sensors as being more compact; living creatures have a higher vibration and a less compact structure, and so on until, as vibrations rise, we arrive at increasingly less compact structures such as gases, air and much more....

Elements such as air and light that vibrate at extremely high speeds are therefore intangible, but this does not jeopardise their existence or their use by human beings. For example, think of the fact that many odourless, tasteless, colourless, impalpable and inaudible gases are in any case harnessed into containers by man and used for his wellbeing. It is on these premises that one can affirm that we constantly make use of elements that, despite their being intangible and non-manifest, are considered to exist as part of our reality.

Cosmologists therefore hypothesized that 90%

of the mass that originally made up the Universe, immediately after the explosion or perhaps up simultaneously with it, impressed such a high vibration as to become imperceptible to its remaining approximately 10%, and therefore also to the human beings who are part of that 10%. They however believe that this mass exists and that it is all around us, at such a high vibration as to be undetectable to our senses.

I fully agree with this theory, because the reality "all around" is often tangible to me in the literal sense of the word.

The solution I was looking for could therefore be found in the cross-point between known and unknown, corporeal and incorporeal, evident and invisible. Only a small problem in the methodological approach remained to be solved: the entire Universe, as a field of research, seemed slightly vast to me, even with all the tools I had at my disposal. However, I have never renounced logic in favour of intuition, nor have I ever done the opposite, and have therefore learnt to adopt the tool that turns out to be most fitting as the case may be. Thus once again , intuition came to my aid with another principle; the law according to which what is contained in the infinitely big is also contained in the infinitely small.

In other words, the Universe is holographic and everything contained in the All is also contained in a part of it.

I thus made use of this principle which, by intuition, I felt could be useful and from there passed on to logic, to reflect and reason on what a dimensionally smaller element that behaves like the Universe might be.

The answer arrived like a shot: the human brain itself. In fact, also of the brain scientists say that only a percentage of between 5 and 10% of its potential is actually used.

So now I was able to narrow down my field of research to a framework that was much closer to me: my brain.

This choice came with several advantages: not the least of which always having the subject of my studies at hand.

3. Brain waves.

How the human brain works has been the object of study for a long time and by various disciplines. One knows that the brain is always active and functioning at all times of the day, even when the body is resting, and therefore not only during the so-called waking hours, but also while sleeping. Scientifically, the brain's activity is expressed by the emission of waves that are referred precisely to as brain waves. These are distinguished by small differences in electric potential which, at a lesser intensity, can also be measured on the surface of the scalp. Their order of magnitude lies within tens of microVolts (1microVolt = uV = 1 millionth of a Volt). Cerebral activity and the consequent emission of waves that oscillate can be measured and visualised by means of a machine that indicates the data on a graphic scheme known as an Electroencephalogram.

Below is a typical example of an EEG chart:

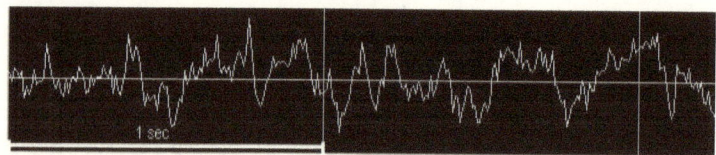

in which 4 types of brain waves can be distinguished, classified according to their frequency, i.e. based on the number of oscillations

per second, which are measured in Hertz.
1 Hz = 1 cycle/sec.

Let us take a look at the four types of waves that the human brain emits:

Beta waves: these have a frequency that ranges between 14 and 30 Hz and are associated with normal waking activities, when the person is concentrated on external stimuli. To human beings, in fact, beta waves are at the basis of the fundamental activities of survival, organising, selecting and assessing the stimuli that arrive from the world around us. For example, while reading these lines your brain is producing beta waves. These then allow for the quickest reaction and the rapid performance of actions.

Alpha waves: these have a frequency that varies between 8 and 14 Hz, and are distinguished by relaxation and meditation states when the mind, calm and receptive, is concentrated on resolving external problems, or on attaining a light meditative state. It is the alpha waves that dominate during moments of introspection, or during those in which concentration is at its sharpest in order to reach a specific goal. They are particularly active upon falling asleep or at the time of awakening, when one is bordering between waking and sleeping. They are typical, for example, of the cerebral activity of those engaged in meditation.

Theta waves: these have a frequency that varies between 4 and 8 Hz, distinguished by the dream state; they are typical of a mind involved in activities of imagination, visualisation, creative inspiration. They tend to be produced during deep meditation, while daydreaming and during the REM phase of sleep, that in which one dreams. During waking activities theta waves are the sign of an intuitive knowledge and of a deeply-rooted imaginative capacity. Generally speaking, they are associated with creativity and with artistic tendencies.

Delta waves: these have a frequency of between 0.5 and 4 Hz and are associated with the deepest psychophysical state of relaxation. The brain waves with the lowest frequency are those that belong to the unconscious mind, of dreamless sleep, of total abandon. In this sense, they are produced during the unconscious processes of self-generation and self-healing.

The transition waves between Alfa and Beta are also called SMR. High frequency waves, starting from approximately 25 Hz, are also called Gamma waves.

This is as much as manuals indicate in relation to the cerebral activities of the brain and its behaviour within the framework of time and space. This is from where I could begin to make my own considerations. After having seen how brain waves work, I understood that the solution I was looking for could be found precisely there, also because, as I said earlier, the way in which reality manifests in the form of particles of matter is only one of the two ways. There is another way, which is that of the wave. This duality of behaviour in the manifestation of contingent reality was used in the Quantum Theory of Fields, which achieves the wave-particle duality by associating particles with *energy quanta* from corresponding wave fields; for example, quanta from the electromagnetic field are associated with photons. In this way, the absolute identity of all the particles of a same type become clearly evident.

On the basis of this duality, I deduced that brain waves are none other than the way the brain was given to create reality and to interact with it under the same form, i.e. waves. By observing the behaviour of the brain and the phases in which the different brain waves are used, I also came to realise the relationship that existed between the latter and the various disciplines I had studied. I understood that, for example, traditional science which makes use of logical-rational reasoning uses primarily beta waves and, in fact, as we saw earlier, these are the waves used for organising, selecting and assessing stimuli that arrive from the world outside a person. We also said that these are the waves that allow for the quickest reaction, which means they are the waves produced by the brain when one accesses the so-called "automatic responses", i.e. the archive of biologically inherited responses, according to some theories, or ones that are learnt from the surrounding environment, according to others. An archive that therefore is in some way extraneous to the person and that is situated in the mesencephalon or midbrain, the part of the brain which also governs emotions. These are therefore the waves used for resolving dissonance, in classic scientific disciplines. In any case beta waves, which are shallow and very frequent, were not the ones I was searching for, in that these are the very ones that manage most of that 5% of the brain's capacity that human beings have become accustomed to using over the past thousands of years. What I wanted was to find access to the remaining 95% of these capabilities and consequently to that part of the Universe that was still unknown. I therefore took into consideration the alpha waves which are less frequent than the beta ones, and therefore deeper. As I said before, scientific studies show that alpha waves dominate during moments of introspection, or during those in which concentration is high in order to reach a precise goal. It is therefore easy to understand that these are the same waves that are used for resolving dissonance in those disciplines that make use of "positive thinking", in those that prescribe mantras or prayers to be recited in which the paced reciting of sounds or words generates a light sort of trance, as well as in other disciplines in which shallow states of meditation are involved. From my own personal experience, I already knew that these methods were good, but I had also been able to see how by interrupting one's conscious attention, the positive processes that had been triggered came to a stop. Therefore, in order to continue to benefit from such practices, one needs to punctually continue with the discipline, and keep ones level of attention high. All aspects which were not in the least fitting to my need for constant change, which badly adapts to repetitiveness of whatever kind or nature. I realised that, to the brain, making conscious use of the alpha waves was the same thing as overlapping a new file of only 2 megabytes onto a more ancient one, already present (in some parts for thousands of years, and in some parts for millions of years) in its memory, and which had a potential equal to hundreds of thousands of terabytes. The new file, taken as a higher vibration, works at the beginning, but as soon as the new vibration falters, is annulled and cancelled by the older and more powerful one which takes the upper hand once again. I knew that all this could only happen because one chose to lay a new and less powerful file over the old file known by the brain, and this triggered the question within me: how and when could this work? How could the file or -if one prefers- the vibration be changed, and the new way of being stabilised for the brain?

I imagined a base structure like, for example, the foundations of a building; to strengthen them and improve them without demolishing the overlying part, you do not place a new structure over it, but you create a new and technologically better part that interacts by integrating with the old one or, more precisely with those parts that are still efficient and useful. In the same way, I was to interact with the old file contained in my brain, in other words remove the parts that were no longer useful to my life and replace them with new ones I had created and that were therefore better suited to my life. But it is easy to realise that this cannot be done using the usual methods and therefore by means of beta and alpha waves: I would have had to work with different waves, slower ones, in order to be able to go deep and work on programs that, often, reside at unconscious levels of our being.

The "right" waves.

4.

I therefore understood that the right waves to achieve this result were the theta and delta waves. We said that according to scientific studies, these waves are active only during the phases of dreaming sleep and deep sleep respectively. One needs to be able to use these waves consciously; this is the real tool I was looking for.

As far as theta waves are concerned, we said earlier that these are sometimes also measured in people during their waking state while engaged in situations of enhanced creativity and are therefore immersed in a sort of creative trance. As we said, during waking activities theta waves are the sign of an intuitive knowledge and of a deeply-rooted imaginative capacities. In general terms, they are associated with creativity and artistic abilities. I had naturally often found myself in that state: while I was intent on designing, when I connected to what I referred to as "the other reality", while I was writing…   I was therefore well acquainted with the conscious use of theta waves, both from personal experience and because I had studied the method by which to use them, coded by a specific discipline during recent years. I understood however that

I needed something different for myself that went beyond what theta waves offered me. In fact, despite them making it possible to interact with the old file, they are not sufficient to replace substantial parts of it, because by nature their activities are based in the midbrain which, as we said, is the site of emotions, but also of the so-called inherited or acquired automatic responses. Because of this it is impossible for these waves to be unaffected by some of the automatic responses deep within, and the changes they work are relative and never absolute. By means of theta waves, one can work changes within the person, in relation to the "customs" of other people, therefore relative. One can for example change the state of sickness into a state of health, but because, with all probability the person, as a result of the dissonance, will have lost the image itself for "state of health", the theta waves will give an idea for the state of health that is based on common thought, and often this relative image is easily cancelled because it gets reabsorbed by the old file. In my experience, I have understood that by means of Theta waves you cannot access what I call everyone's Personal Vibration – which I will discuss in detail further on – within which the absolutely best state of health for that particular person can be found. You can only thus project an image of health that corresponds to statistical values. If you think about this and transfer it into people's behavioural field, one can only project stereotyped images: think of the image of good, of joy, of happiness, etc., all based on clichés and not on personal awareness. Like this, it is as though one removed one kind of scheme from the person and replaced it with another derived, in any case, from value judgments which, in my way of thinking, prevents individuals from being just that, and ends up instead making people who perhaps are happy, but who are all the same.

For all these reasons, despite it being optimal with respect to my initial starting point, the result achievable by means of these waves was not good enough for me, given that I was now determined to find the definitive solution and thereafter pass on to doing something else.

I acknowledged the fact that what I was looking for could only begin from the delta waves. By definition, all we know is that they were produced in the area of the front lobes, known as the Silence Zone. A zone considered to be inaccessible on a conscious level.

The two years I had spent only during research had been used to find the way to gain conscious access to them in order to be able to use them.

5. The deep brainwaves.

Do not ask me to prove what I am about to affirm by means through laboratory experiments because I would not know how to do so, but what I do know – as a result of numerous practical experiments – is that by using these waves people's lives can change profoundly to go back to living their lives in perfect balance with the entire Universe, as is their very nature. Those who have experimented with me during the past two years can attest to this, because they have seen their lives change radically. They have chosen the quantum reality that they wanted to live and are now in the harmony of the All. From beings in need of help they have become beings who can help, because they have regained charge of their lives.

As far as I am concerned, I will only try and explain how what I like to define as being a **way** works, amongst the many ways possible, in order to take a quantum leap. A way by which we can change our reality and follow new quantum possibilities to fulfil our life.

I will talk about my experience andt about the message I received on the "law of dimensions", that I was able to understand thanks to the use of my innate ability to connect with parts of the Universe in which it is also possible to also learn about teachings as yet totally unknown to us. Or, if you prefer to see it from a biological point of view, I will tell you about the understanding I have reached by reactivating the memories that have always been present in my cells yet suppressed for millennia. This demonstrates that what is contained in what I previously referred to as the ancient file is still useful in part; it still in fact contains information that is extremely precious for human beings, it just depends on what use they make of it.

The Law of Dimensions works by means of the conscious use of delta waves and is used to manifest one's reality on all the dimensional Planes of Existence, which I will talk about in detail in the next chapter.

By means of the Law of Dimensions and the consequent use of delta waves, we can create our reality on all planes, both in our dimension, and in others.

What I know is that while theta waves are propagated in mediated resonance with the Universe, and therefore do not have an immediate effect on the reality in which we are immersed bodily and materially, delta waves have a multiple characteristic, which is that they are propagated both in immediate resonance with the Universe with respect to our material and bodily reality and to all the other realities in which human beings exist, and in programmed resonance with the Universe. This means that through them we are able to interact both on the Third Plane of Existence, which is the one in which the Earth and the human beings who live on it exist at the current point in time, as well as on the other twenty-two dimensional Planes of Existence, in which every human being exists despite their not being present there physically. This is because the Law of Dimensions transcends the laws of the Third Plane of Existence.

The statement I just made has implications of enormous reach. In fact, when one thinks of its effects, this Law provides access to immortality, to immunity, to infinity and to immateriality, all of which are aspects of other Planes of Existence, as we will see in detail in the next chapter. Some parts of the Law of Dimensions were also known to the Ancient Peoples – about whom I will speak in detail in the next chapter – in times very distant from ours. However, I know now that this law has never before been known on Earth, in its entirety and in this form. In fact, when the Ancient Peoples used this law, they did so on a different dimensional plane; in other words, the person transferred his or her being to another Plane of Existence and from there was able to use the law. This way of working never made it possible for the Earth to make full use of the Delta potential. Now, the possibility of making full use of this law on our dimensional plane takes on new meaning, and the opportunities its use grants can easily be imagined. For example, through it access can be gained to the laws regarding our convention on time. Because it is a convention, time is an element that is strictly connected to the dimensional plane of human existence, and therefore has no influence over the Law in question. It follows then that one of the corollaries of the Law is immortality of the body, because the laws to which until now it has been subject are those connected with the conventions of linear time which establish its cycle and therefore also its ageing and death.

Moreover, the Delta Law allows access to a different use of matter, because this too is connected to a human convention which is that of space. In this field, already in the past there have been accounts of matter being used in a different way; think for example of some of the martial arts in which one can walk on air even if only for an extremely short span of time, not to mention He who walked on water...

Different ways of using matter, therefore, until now called miraculous or almost, whereas knowledge of the Law of Dimensions makes it possible to dematerialise the body and anything else present in the current dimensional plane and materialise it in any other point of the Universal Vibration (a detailed explanation also of this will be given in the next chapter).

Given the enormity of this information, it came naturally to me to ask the Universe to comprehend what the ultimate goal of the use of this Law was, and the answer was as surprising as it was simple: "to bring wellbeing to the world".

The Law of Delta.

As we said earlier, the Law of Dimensions transcends the other laws of physics present in the human bodily plane of existence and is therefore not subject to them; it thus overcomes the law of compensation that belongs to this dimension and does not alter the equilibriums of the Earth when it is used. This means that no problems are created in any part of the globe or other dimensions. The Law was given to us for our utmost good and for the utmost good of the entire Universe. Even if one is sceptical about it, it can be used anyway, because it is a Law and it works.

In order to use this Law, the front lobes must be released, and subsequently the so-called "Silence Zone" freed. Releasing this Zone of the brain signifies being able to use the delta waves that are formed here in a conscious manner and in a waking state.

After having released them, through practice we arrive at mastering the conscious use of Delta waves and through them we move on to the active phase, which consists of localising the "map" of all the various aspects of our life. This is necessary in order to clearly identify where, how and when to make the things that we want and that are useful in fulfilling our needs to happen. Knowing the map makes it very easy to progress, because we can go and create reality in the exact point of the photon trail -that corresponds to the quantum possibility we are living- and that we most want to change. More precisely, if we look at it from another point of view, we take the quantum leap by directing ourselves exactly to the photonic trail we are most interested in experiencing as our new possibility.

This was my first approach to what I called the Law of Delta. I later arrived at other conclusions and a new understanding.

What I know now is that Delta waves are very slow and therefore broad and deep and that they have a very high vibration, such as to make it possible to interact with everything and such as to make it possible to take the quantum leap. The most important thing is that through them we can access our own maximum vibration, in other words 90-95% of the Universe or, similarly, of the unused potential of our brain, and acquire the conscious image of it.

As a result, when we change our life, we do not enter other convictions or beliefs taken from outside ourselves, but we can project that which is the absolute best for ourselves. This means that the Delta waves no longer give the relative image of a part, but the absolute image of the All.

In other words, they make it possible to access the circular time of the Universe, moving away from our linear time and allowing us to continuously take quantum leaps.

Delta gives access to what I call the Personal Vibration, about which all now is that it consists of all the possible space-time crossings I will say for each individual person; more detailed explanations will follow shortly.

Have you any idea of the enormity this implies to human life?

It means that we can consciously choose the path we want to travel in our life, in all sectors of our life, in every small nuance of our daily life.

It shows that we can choose the situation we want to experience from the view point of physical, financial, emotional, intellectual, spiritual wellbeing….

It means that, by raising our vibrations, we have more tools with which to create our reality. Therefore, we can harmonise with the All, with the Universe, and draw whatever we need from it, at whatever level. Lastly, we can choose the lane in which to continue our life. It is clear that in order to do so one must vibrate at the same speed as the photons that make up that reality. That is why I say that in order to make that change, we need to manage to vibrate in Delta, the vibration of which is equal to that of light.

Moreover, since everything that concerns human beings belongs to a specific "Plane of Existence", with delta waves you can decide what to manifest and when. Through them you can access all Planes unconditionally and to interact with all Energies. This establishes another thin line of distinction between theta waves and delta waves in creating one's reality. In fact, we said before that the main difference lies in the fact that theta waves are propagated in mediated resonance with the Universe, whereas delta waves are propagated in immediate and programmed resonance with the Universe. Now we can say that this signifies that the former are propagated in the Universe through a means, through something that supports them, whereas delta waves propagate anyway, regardless of everything. Therefore theta waves propagate in matter, without the substrate of which they could not exist, while delta waves exist regardless of it. This is also why delta waves exist on all dimensional Planes of Existence, while the theta ones only exist in those Planes in which the vibration is lower.

And again, what does it mean when we say that delta waves are propagated in programmed resonance with the Universe?

It means that, because the effect requested from the waves takes place in the quantum reality in which the space-time convention does not exist, we can program it so that it materialises at a certain time and in a certain space according to the rules of the dimensional reality in which the Earth finds itself at the current time. The immediate consequence is that if used consciously, delta waves give access to immunity, immortality, infinity, transcendence of the laws regarding the three dimensions, immateriality...

The only law regarding delta waves is called the *"The Law of Dimensions"* and, as I said earlier, this has never before been known on Earth in this form.

It is not subject to any of the laws of physics, nor does it disturb anything that is subject to them, but is simply capable of interfacing with the All, only changing that which needs to be changed, without any side effects.

It simply Is.

Anything can be done with it and soon it will be accessible to everyone. In the following chapters I will describe the method that I have created and by which it is possible for anyone to understand where they are at this point in their lives, where they want to go and, above all, how to get there.

The following information therefore explains how to achieve all this and prepare oneself to be activated in delta.

CHAPTER II

THE CURRENT PERIOD IN TIME

After having illustrated the various phases that in time led me to search for a new path towards change and the quantum leap, I will now talk about the relationship between the search for a new path of knowledge and the historical period in which it falls, highlighting how the objective existence in time of the path is the "fruit of its times", why this is so and what the ultimate goal of all this may be.

Moreover, we find ourselves facing the important question that has always tormented every human researcher throughout history: how far can one push oneself along the path for Knowledge? At what point is it best to stop? Where is the forbidden limit beyond which one cannot go?

One will realise that the sensation of "surpassing limits and boundaries" is only an impressed memory, a conviction, a belief, and therefore factitious in itself.

It is man's fear of "eternal hell". Upon closer study, we can see that it is precisely in the moment of doubt, in the instant in which we judge ourselves and condemn ourselves for our conceitedness that we really could become conceited. If there were, a point beyond which we really could become conceited, this point would coincide precisely with the instant in which we interrupt our path of knowledge for fear of exceeding its limits.

If human beings follow a path in their evolution, it means it belongs to the quantum possibilities available to them. Therefore, whatever the path you choose to follow, you are actually only walking a path that already exists in the Universe and that, as such, is accessible to human beings. And so who has the right to decide for themselves or for others when and how they must stop themselves from doing the will of the Universe? True conceit -if it did exist- would consist in interrupting the path of Knowledge.

1. Sense-project of the research.

First of all, you must understand that anything, whatever it is, can only exist in the Universe up until it makes sense for it to exist.

A house, a relationship, a friendship, a human being, a life... everything can only exist up until it makes sense for it to exist.

The sense of something is the reason for existing, the raison d'être, that we give to that thing.

Its sense arises from a need.

In detail, we can schematise how this works as follows: you feel a need, you make a project, then you create something that has the sense of satisfying that particular need.

A practical example that explains this concept is as follows: in a certain area of a town, the need arises for a place for reading and for culture, in which to exchange ideas, etc. This will lead to a project being drawn up and a suitable building built that responds to that need, a library for example. The sense for that library to exist will therefore be to fulfil the particular need that triggered the project behind it.

Then, one day, passing through the same area of the city, we see the library has been turned into a residential unit. This means that for that area of town, the library structure had terminated its sense-project, one of study and culture and socialising, and has taken on a new role, one of providing housing in order to meet the new need of the people who reside therein.

Because this concept applies for all things that are created, the study and method I will proceed to explain has its own particular and precise reason for existing in the Universe at this precise moment in time.

During my studies, I have always tried to be aware of what was going on in the current period in time, in order to sense towards what quantum possibilities of evolution the life of human beings today could lead, both in an immediate future and in a more distant one.

I thus understood that there is a particular quantum possibility that grants every human being the capacity to evolve for themselves, according to their own biological needs and, simultaneously, precisely by doing so, to also enable those around them to evolve.

All this made great sense to me, so much so as to induce me to create this method, which was born as a powerful instrument to evolve and to make evolve according to that particular quantum possibility.

Because I know that the further ahead in time you project the goal to be reached, the longer the object of creation will have to last -and thus the longer the time it will make sense for it exist-, I developed this tool in such a way as to give it the maximum adaptability, free from schemes and conventions, and therefore in constant evolution in and of itself.

In brief I can say that the sense-project I gave to this part of my path is to help people take their quantum leap within the framework of the possibilities that the Universe places at their disposal in order to evolve so that they, in their turn,

can contribute to making others evolve. I have chosen to apply my method to this precise quantum possibility because, being part of linear time, I know for sure that in the particular period of time in which we live, something very important and very nice is happening to our planet.

The time has come to accomplish a momentous passage that will lead to a great evolution of the human being, and each of us can contribute towards it.

We are not alone, but like all things that take place in our Universe, everything has already been arranged in the best possible way.

That is why already over the past several years there are numerous scholars who are working, in various parts of the world. Through study, research and theories they are showing the masses the way to be followed, so that everything takes place in the best of ways.

The task of these illuminated people is that of bringing awareness to as many people as possible, to help them in what is referred to as "the passage".

Many have studied the ancient texts of various cultures which contain indications as to what will take place in this era, and are spreading this information through great and substantial work carried out worldwide. Others have chosen to help in the passage by using abilities inherent in human nature, but forgotten for thousands of years. Still others again have studied and refined techniques to restore balance to people's minds and bodies.

Some do this from a spiritual point of view, others from a bodily point of view.

It is of no importance what method of work is

chosen; the important thing is that everyone is working towards helping the world in this momentous passage.

Every technique and every teaching is equally important and is developed to a different degree of depth, because that is what the World needs right now.

Not everyone has the same degree of preparation and perception; there is therefore a discipline suited for everyone.

And so -in order to evolve- everyone will choose the way they feel closest to them; they will follow that path according to their own time and possibilities.

I personally only know a part of what will happen in the world after the passage, but I know for sure that if people are given the highest degree of adaptability, they will be prepared to live any change well.

That is the path of evolution that I have chosen to travel: help people live their current life well and prepare them to live any future situation well.

To live the current life well, we must be aware of what is happening on Earth and all around us at this particular point in time.

2. Old and new grid.

Already approximately twenty years ago, geologists and scientists of other disciplines realised, thanks to their studies and measuring instruments, that the rotation of the earth around its axis is starting to slow down (this implies a change in the magnetism of the Earth itself) and that, at the same time, the frequency of the Earth, i.e. what is referred to as the "throb" or "pulsation", is increasing,.

According to the studies conducted and those still underway, it seems that this phenomenon will lead to a moment in time when the rotation will reach its lowest level and the pulse its highest one.

In a hypothetical graph, the point of intersection of these two trend lines is called the Zero Point.

In the instant in which the Earth reaches the Zero Point, substantial changes will occur, the most evident of which will probably be the inversion of the direction of rotation of the Earth on its axis.

The moment in time that ancient texts have already referred to as "the day in which the Sun rose twice" will happen once again.

According to scholars of the field, this moment has already taken place various times during the existence of this Universe. It seems that the last time was only approximately three thousand five hundred years ago, and various historical evidence exists to confirm this. Therefore, what is about to happen or, rather, what is already happening, is something that has already happened at other times during the history of the Earth, therefore there is nothing catastrophic about it. What has already been taking place for the past few decades is simply a loosening of the electromagnetic "grid" that envelopes the Earth.

The production of an electromagnetic energy field is caused by the vibration of the Earth itself or, in other words, by its vitality. In fact, every living being produces an electromagnetic energy field that can be measured around its body. The energy field of the Earth is kept together - like an imaginary grid with a very tight mesh - by its speed of rotation around its axis. Over the past few decades, this slowing down in the rotation has given rise to a loosening of the cohesion between the mesh of the electromagnetic grid, allowing for a greater amount of information between the Earth and the remaining part of the Universe to filter through. Just like in an imaginary grid, the mesh has become wider and continues to do so, allowing human beings to acquire information that for thousands of years has been precluded from them.

This information has always resided in their DNA, but has remained latent for a very long time, because it was necessary that it remained so. In this particular moment in time, also thanks to conjunctures of a geological and astronomical nature, access is given to everything that the human being is.

This is the time of awakening for everything that up until now has remained suppressed. Access will remain "open" until the Earth inverts its sense of rotation, acquiring greater speed and thus reforming a new grid. If we work well, the new grid will contain information of infinite joy, beauty, harmony, love and life.

This is what I refer to as the passage from the old grid to the new one. What bad or terrible things can there be to all of this? There will simply be changes to which human beings will have to adapt, but everything has already been arranged so that this takes place in the best possible of ways. For instance, it has been noted that an increasingly large number of people born during the last twenty years have "peculiarities" of a biological nature, such as the so-called "double brain". This definition is used to indicate people in whom the cerebral connections between the two hemispheres of the brain are in much greater number than those in most people. This characteristic makes of these individuals people who are gifted with the ability to do at least two things at a time and who therefore require half the time compared to others to perform particularly complex tasks, as well as with the ability to maintain maximum concentration, connect with vaster parts of the Universe compared to strictly material ones, easily heal themselves, never have serious illnesses, and still more. All this because the faculty to use the two hemispheres of the brain at the same time is constantly active. These people therefore adapt immediately to all events of whatever nature. Obviously, what I have just described is a double-brained person in perfect balance with himself and, therefore, free of dissonance.

Over the past twenty years, the percentage of people who have this characteristic and who have already been identified and studied in psychobiology for some time now, seems to have passed from 3% to 5% of the population worldwide. Similarly, scholars of the sector have noticed a rise in the percentage of people who have different characteristics in the field of abilities we could call "spiritual"; people who in other words can spontaneously activate self-healing abilities, or who have clairvoyant or clairudience powers... . This second type of person has been referred to in time with terms like *Indigo Child, Rainbow Child, Crystal Child*... I personally like to simply call them all "Crystals", because at this point in time they have the same transparency and fragility as crystal.

I am firmly convinced that all these people are already born ready for the evolution and passage that the world is preparing itself for. I am especially sure that the first type of person - the so called "double brains" - include those who are capable of transmitting the biological information of maximum adaptability. These people in fact are already born "prepared" in such a way that adaptability and ductility in all fields and in all situations are their essential characteristics. Even if they wanted to, they could not live differently.

What I know is that after the passage, on Earth there will be a different way of perceiving and conceiving time and space and, because of their biological structure, these double-brained people will know how to do it in all sectors of their life. They are already naturally prepared for life in circular time (the concept of which will be discussed in greater detail in the next paragraph).

"Crystals" too have the innate ability to live in circular time, but limited to the "gift" that distinguishes them.

This makes "crystals" extremely fragile. It is in fact very difficult for them to live within the framework of linear time, because when they are born they are biologically prepared to live in circular time, but they lack the characteristic of maximum adaptability that "double-brained" people have that would enable them to live well in both situations.

In fact, whereas the "double-brains" are born perfectly suited for life during the passage from the old space-time grid to the new one and are therefore able to live well both in linear time and in the multi-dimension - therefore in circular time - "crystals", on the other hand, are born to live in the new space-time grid, in which multi-dimensionality already exists and sequentiality no longer exists.

With these characteristics, in their daily life during the current time of transition, they experience a discrepancy between that which day-to-day life with its rhythms and timings requires of them, and their innate inner time. This discrepancy often leads them to having what most people consider anti-social behaviour. In particular, this happens to crystal children in institutionalised places, such as schools. All this makes it easy to understand that the ultimate goal for achieving maximum adaptation and avoiding situations of people who are poised between two realities is to teach those who are not born biologically prepared for maximum adaptation, to be in balance.

This can be done. People can be taught to live in both grids, as needed, and to manage their nature in the best possible way for themselves and for others.

Basically, this involves helping as many people as possible who are already born "prepared" to activate their innate gifts. A result that is easy to achieve, by:

Restoring balance to every sector of one's life. This result is achieved by changing the stereotyped image the person has of themselves, with their own absolute best image, taken directly from their Personal Vibration. For example, in the case of "crystals", this would be to teach them to be in balance with themselves, according to their innate needs, which are ones for structure

Reactivating their latent cellular memories, which allows everyone to use all the potential inherent in their genetic heritage.

Showing people the way to attain maximum elasticity and the ability to use all the instruments possible that the Universe makes available to them.

Teaching meditation in delta, by which each one can, on their own, accomplish all the steps desired, endlessly.

CHAPTER III

OTHER DIMENSIONS

Once it has been understood how brain waves work, the differences between them, the prerogatives inherent to each type of wave, the quantum possibility chosen and the sense of this disclosure, it becomes necessary to enter into further detail in relation to particular issues to which reference is often made herein.

To this end, this chapter will introduce concepts such as circular and linear time, ancient weave, multidimensionality, dimensional Planes of Existence, image, Personal Vibration, Universal Vibration. . .

These subjects will be discussed summarily, and only in as far as is useful for understanding the main subject of this book, which aims fundamentally at expounding the utility of the Law of Dimensions and positioning it within the current historical moment, literature and social environment, so as to comprehend its potential and interactivity.

For a more detailed study of these subjects, please refer to other writings by the author.

1. Linear and circular time.

This definition as such is not correct, because it would be better to say linear time and absence of time. The concept of circular time, however, is more comprehensible to the human mind, and so it will be to this that we refer in distinguishing between the now and the then.

As already mentioned when the concepts of "old grid" and "new grid" were introduced, the Earth is preparing itself for the dimensional passage that will carry it from the third (spatial) dimension to multidimensionality. This means it will pass from the space-time dimension we have known for the past thousands of years to a situation in which time and consequently space is absent or, (which is the same thing), to a situation of Infinite Time and Infinite Space. The expansion of human awareness will therefore necessarily pass through "modulating the brain to the concept of Infinite".

This involves bringing the information of "Infinite", complete in its every part, to the brain. This necessity arises in relation to the fact that in the new quantum reality we must be able to conceive of Infinite place, Infinite space, Infinite change, Infinite places, Infinite worlds, Infinite interactions... in other words conceive of Infinite life in the world.

Right now, in the human brain, "Infinite" only exists as an idea and, because it is absent at a conceptual level, finite concepts clash within in that recall images of boredom and pointlessness of infinite life. This happens because life extended to the Infinite is, now, perceived as repetitive and therefore useless. By introducing the concept of Infinite and, especially, through the awareness of this knowledge, (acquired by introducing the image of it into the brain), we can expand our horizons and start to abandon the image of finite that has till now permeated human life on Earth, giving rise to duality and contrast.

Up until now, the human brain has only thought about the Infinite, in other words it has had it as a thought, but has never taken it to the midbrain and to the brainstem to make it become part of it. The absence of the image of Infinite in the human brain is due to the presence of the image - acquired and not innate - of linear time. It is the linearity of time that carries within it the implication of finite.

Linear time, in fact, corresponds to the image of a line that has a point of origin and a point of arrival and that exists thanks to a sequence of points; the line itself is derived from joining these points.

The timeline in and of itself therefore contains the idea of advancement and consequentiality. In it, everything that exists follows step by step and necessarily has a beginning, an evolution and an end. Everything that has characterised the life of human beings on Earth during the past millennia has this kind of concatenation. Thus, on the basis of this convention, human beings are born, grow up, age and then die, according to a fixed sequence of interdependency of points, i.e. according to a linear kind of sequence. In the same way, things, situations, animals, and everything else that belongs to the same human plane of existence follows temporal linearity. That is why each time we "talk" to the human brain, we must follow this same linearity and proceed step by step. Simply use this book as a practical example of what I am asserting. In it, it is essential to place one word "after" another, construct sentences so that from one concept we can proceed to the next; then from one subject to the next... in a progressive sequence, so that the final structure obtained is a book that has a beginning and an end and that can in and of itself be considered finished, or "finite". The human brain is so used to doing this that in the precise instant in which it conceives the idea of creating something, it has already decided that it will have a beginning, a progression and an end.

That is what always happens and this will become clearer in the paragraph dedicated to the "sense-project" in which it is explained how the loss of the original sense of anything determines its ending. It is easy to comprehend that the reason for which this can take place is the existence of the concept of linear time.

Temporal linearity is, therefore, what has conditioned the unfolding of life on Earth during the past thousands of years. In the immediate future, however, after the passage to the new grid and therefore to multidimensionality, the concept of linear time will lose its sense for existing, and human beings will have to live in "circular time".

What I mean by "circular" is the simultaneous nature of everything.

Everything will take place at the same time, and the need for sequentiality will no longer exist. Human beings will use all of their potential fully and, for example, will be able to understand entire discourses at the same time, without having to listen to the sequence of words pronounced one by one.

In a dimension in which everything happens simultaneously and everything is in the present Infinite, human beings will adapt their way of perceiving in order to be in infinite simultaneousness. They are already capable of doing so. The Universe, Infinite, Immeasurable and Astonishing, has already provided human beings with the ability to learn and understand in circular time. Everybody, in fact, has the inborn capacity to understand music and art. These are two disciplines that only exist in circular time and that men who are deemed to be extraordinary - musicians and great artists of all times - have managed to comprehend and bring into linear time. That is what happens with music: the more the notes and the instruments played at the same time, the more the enjoyment for the human brain; with art, for example pictorial art, where the presence of infinite colours mixed and blended harmoniously strikes the human imagination creating emotion; with poetry, where the musicality of the verse enchants the listener's soul even when it does not follow a logical sequence or when it closes hermetically in the contraction of a phrase, dropping the subject or the verb in the most total grammatical anarchy; with literature, and with anything else that can be defined as Art. The human brain is capable of understanding this at a profound level. The artist -a privileged being in strict connection with universal knowledge- grasps the infinite harmony in circular time and brings it -with his innate ability- into linear time, in which everyone is capable of understanding it.

Human beings, with no exception, are already born prepared to live in circular time, and always have been.

At this time in which we are shifting from one grid to another, however, we can see that some people are born completely prepared for multidimensional life and are incapable of mediating between linear time and circular time. Often even great artists are unable to live in the linear world outside of their art. We could say that these people live perpetually in a musical dimension which, however gratifying and harmonious, clashes with the surrounding reality that is still immersed in temporal linearity.

That is why it becomes indispensable to first and foremost accept the old structure, -bound to finite concepts, linear sequence, time and space- that still remains in human reality. Only after the old structure has been acknowledged and accepted can one proceed to integrate it, teaching those who are in linear time to also live well in circular time, and teaching "crystals" to also live well in linear time. This result can be achieved by activating, particularly in the first group, the memories that allow them to perceive circularity through music and art, whereas with the second group it will be necessary to interact with the genetic memories inherited from their forefathers who lived who lived in linear time. Simply by activating these memories, it becomes possible to enable everyone to accomplish the passage of grids, without trauma or maladjustment, and be capable of managing any of the situations that may arise on Earth following the shift to multidimensionality.

The ancient weave

2.

The "ancient weave" corresponds to what up until now has also been referred to as the "old grid".

This is the old Vibration that belongs to everyone and that is closely interconnected with the Earth. We said that because of the slowing down of the rotation of the Earth around its axis, the electromagnetic energy grid produced by the Earth and by all the living beings present on it is widening, because the speed of rotation is no longer such as to hold it together with the same strength as before.

We also said that at least one of the components of the old electromagnetic grid, i.e. the passage of time, is imperceptibly changing, precisely because of the lack of cohesion. These elements of change are also changing people's feelings and, especially, the way that they express these feelings themselves, because feelings are none other than electromagnetic energy produced by chemical exchanges inside human cells. From all this it is easy to deduce that the change in feelings and therefore of chemical exchanges is also changing the human DNA.

By means of the delta awareness path, we can - on a personal level - speed up this natural "process", arrivingatconsciouslychangingourelectromagnetic vibration and therefore our own reality.

This means changing our DNA and therefore our materiality.

Forallhumanbeings,thisisthetimetofreeourselves entirely and close the emotional sequences left open in the "old grid" for once and for all.

By shifting to the vibration attainable by means of the Law of Delta, the chemistry inside us also changes and, as a consequence, so does our electromagnetic vibration. This kind of change leads us to naturally detach from the ancient weave, so much so as to be able to see it with our naked eye. Once our vibration has been raised, we can look around and see the "ancient weave", which looks like the microscope enlargement of live, moving bodies, suspended in the air. They appear very far apart from each other and positioned exactly as though they were the residue of a line that originally created a very vast grid. In order to forever abandon the old grid or, in other words, the old conflicts, the old behaviour, the old habits or, simply, the old life, one needs to free oneself of these residues.

When the old grid is removed completely, the quantum leap will have been completed and you will be in the new reality created, and this can neither stop nor be stopped, whatever you do from that moment on.

3. The Universe.

At this point, it is necessary to understand the structure of the Universe in which human beings exist.

In the study of physics, already at the beginning of the last century, a theory emerged, known as the String Theory.

This theory arose from the need to reconcile the principles of Quantum Mechanics with those of Special Relativity. At a certain point of the research on the so-called "impact processes" between particles of matter, (which play a fundamental role both from an experimental and from a theoretical point of view in the physics of elementary particles and are the primary means for studying the interaction between them), Quantum Mechanics introduced new elements. The latter in fact assigns oscillating properties, as well as corpuscular ones, to the two types of particles called fermions and bosons respectively. Very briefly, one can say that this is the main reason for which since the end of the 1920's the problem of systematically reconciling these new principles with Einstein's Special Relativity has arisen with growing insistence.

The Quantum Field Theory succeeds in this goal, which attains the wave-particle duality by associating particles with *energy quanta* from corresponding wave fields; for example, photons are associated with quanta from the electromagnetic field. Like this, the absolute identity of all the particles of the same type becomes clear.

Subsequent experiments have shown how the few particles that make up matter -electrons, protons and neutrons- are accompanied by a multitude of other particles, the majority of which are unstable. From the '30's onwards, several attempts have been made to establish a theory for all the elementary particles.

The String Theory falls under this research. This theory seems to be the link that scientists are looking for but, in order for it to consist, it calls for the existence of a good *twenty-five spatial dimensions*, instead of the three we have in our daily experience. In interpreting the Theory of Strings as a basis for uniting gravity with the other fundamental interactions, it therefore became necessary to link it to the human perception of a Universe with three dimensions of space. By means of subsequent tests, the conclusion was reached whereby the universe could possibly contain some dimensions that had crystallised on a microscopic scale during the initial instants of expansion of the Cosmos, but both General Relativity and the String Theory are apparently unable to explain why this happened. Now, scientific research in this field is directed precisely towards reaching a better understanding of whether or not there actually are other dimensions in the Universe and why they crystallized on microscopic scales.

Personally, not having had any specific scientific training, I do not find it difficult to consider the existence of an additional twenty-two dimensional Planes other than those we are familiar with through our physicality. And therefore the very reasons that prevent scientists from applying the String Theory, i.e. the co-existence of twenty-five spatial dimensions and the different vibration of matter that renders an infinite number of molecules unstable other than the three that make up matter to the contrary is, in my view, reality itself.

Leaving scientists to discover why dimensions other than the ones we know are contained in the Universe, I prefer to instead turn my attention to exploring the possibilities and ways of using all the dimensional planes of Existence, to the benefit of the practical life of human beings.

From the path of knowledge travelled till now, I am in fact certain that at least twenty spatial dimensions or dimensional Planes of Existence exist in our Universe. These are twenty different Planes in which all things and all beings, exist in a different way. Therefore if, for example, in the three most commonly known planes, a human being exists on a Bodily, Sexual, Intellectual and Emotional level, on other dimensional Planes he or she will exist on a Spiritual-Emotional level, or on a Spiritual-Intellectual level, etc. One thing is for sure and that is that on the level of delta waves human beings exist on all dimensional planes, with no exception.

That is why these Planes can be called Dimensional Planes of Existence.

In general, one can say that the Universe in which we find ourselves consists of twenty Planes of Existence, only three of which are situated in the space-time convention and are perceptible through our five senses (with the use of beta and alpha brain waves), whereas another seventeen can be perceived and reached by means of the deeper brain waves and, more precisely, up to the seventh dimensional Plane of Existence by means of theta waves and from the eighth to the twentieth dimensional Plane of Existence, by means of delta waves.

The passage of the "material" part of the first three planes of existence from three-dimensionality to multi-dimensionality therefore implies a rise in the vibration of matter which, in particular, will cause the third Plane of Existence to evolve towards other dimensional Planes, without it being necessary to abandon one's physical bodies.

The Earth will therefore shift to the maximum vibration in which all the other Planes exist, thus passing to multidimensional life.

Here, all the dimensions will be tangible simultaneously, and every being will instantaneously be perceived and will instantly perceive the All.

The All is composed of infinite Universes which, according to the degree of evolution reached, are divided into an ever smaller number of dimensional planes.

Because the All is One, all the other Universes as well as ours will be affected by the dimensional passage of the Earth

4. The Dimensional Planes of Existence.

It is wise to specify that when we talk about dimensions and dimensional Planes, indicating them with progressive numbers from the third onwards, our intention is that of giving the human brain an idea of evolution, advancement and change. By means of the numbers, in fact, the brain paces a linear kind of rhythm, but it is obvious that once linear time has been abandoned, one is automatically in multidimensionality, and therefore saying fifth or twentieth actually has no sense. This distinction, therefore, is only maintained to facilitate understanding and to distinguish the specific prerogatives of each of the various Planes.

A summary of the main characteristics of the various Planes of Existence follows.

VII Dimensional Plane: here one can experience an immediate awareness of the All. Such a state of awareness makes it possible to instantly create reality in the bodily, or three-dimensional, plane of existence. Everything you want to change in your life can be created in the seventh Plane of Existence. In this dimension, one has the potential of balancing on a bodily and material level, for oneself and for others. However, this plane is still linked to and moulded by the laws of space and time of the human dimension, and therefore the interaction manifests according to preconceived times and ways in the mind of the healed person.

VIII Dimensional Plane of Existence: here it is possible to create healing at all levels of Being, forever. It is given by the strength of weakness, which is infinite energy in motion. It represents what over the centuries has been referred to as the Sacred Feminine, the void made and understood in its energy in motion. The Energy of the VIII plane can be seen as infinite points of light made of little spheres, constituting the aggregating element of the energy bodies of all Beings.

IX Plane of Existence: the infinite perception of the Divine Being, which is awareness of the All, can be perceived here.

X Plane of Existence: this represents the annulment of the All in the divine form. In it the form of all things or beings can be seen, that invariably corresponds to energy that is pure, luminous and free of any aggregation.

XI Plane of Existence: this is where the constant and infinite regeneration of Light, in other words of Energy, takes place.

XII Plane of Existence: this is the door that gives access to being Everything that Is. Before we cross the Sacred Door, we can work at the maximum level of transformational Energy, whilst still being aware of the three-dimensional Plane of Existence. Once we cross the Door, we immerse ourselves in the World of the Soul, in which we are the essence itself of the thing we are working on, and this is aware only of itself, nothing else matters to it.

Once we cross the Sacred Door of the XII Plane of Existence, therefore, we enter the place of the Soul, in which we understand that our own soul is also Anima Mundi or the World Soul, and that this has its own different knowledge. Its knowledge is being Everything, all things, all creatures, all atoms, and experiencing that thing with awareness of oneself, and ultimately becoming that thing. It is not the same thing as feeling part of the all, but it is being every single thing, because everyone is Everything. However, when one is that thing, one loses "external" perception, and therefore it is better to work on the threshold, i.e. in the XII dimensional Plane of Existence, in which we still also maintain the perception of reality as outside of ourselves; this makes it possible to perform the balancing by means of the awareness of the All and the maximum form of adaptability.

Sometimes it is necessary to go to the World of the Soul, in order to really understand what a cell or a certain part of the body is feeling, and ask it how it wants to be re-harmonised...

The XIII dimensional Plane of Existence is therefore made up of the Anima Mundi which, in many traditions, is known as the Akasha. Here one can hear the Soul of the World speaking. It is wise, it knows everything that is; it has been and will belong to Everything and to Everyone. Here it is possible to work with every single Being of the human world both to learn from it, and to help it. Everything can be healed from this Plane: earth, fire, air, water, human and other beings, all Beings of Light. Here, balance can be restored to everything that has lost it.

In the XIV dimensional Plane of Existence, the Great Wonder of Life can be understood. In it, all things flow with the movement of a river of white light that cradles the traveller upon his arrival with soft, light waves. It is easy to abandon oneself to it and the slow waves relax one immediately, cradling one and accompanying one to an even deeper part of one's Being. To the river, you can forever abandon all thoughts, angst, anxiety or worries that still lie hidden or latent within you. All unpleasant thoughts regarding any level of human life can be abandoned definitively and left there. The depth to which we can reach within ourselves from this Plane of Existence is such that the body often shows its effects through a sense of nausea. We need to learn to throw everything found in the depth of our being into the river of quantum possibilities, because the river of photons that can be seen in this dimension is none other than Lethe, the River of Oblivion of the Soul.

In the XV dimensional plane in which human beings can exist, one finds the Passionate Love of the Universe. This is equivalent to the materialisation and dematerialisation of everything that Is. The light is white and is shaped like a star. The fringings of its rays have golden nuances. In this dimension, one feels one is entering the heart of white light of the star. Within it the dematerialisation of one's being takes place, which merges with the light becoming an agglomerate of shining photons. Even the outline of the human shape is lost in this dimension, so that the Passionate Love of the Universe reaches those who are capable of perceiving it, shaping us in its image and likeness. Life itself flows in those who know how to consciously reach this dimension. The human soul is revived. Whether it was dormant or lifeless, or had never been in the body, it comes to life.

The XVI dimensional Plane of Existence is given by Unknown Love. Even here a star of brilliant white light, with very fine light blue veins and transparent threads, can be seen. The impression is that the air made of light moves towards the beings with an undulating motion. Tiny waves of light that flow as though to form a vertical membrane in relation to all beings. This reality actually comes from a dimensional membrane that leads to the V dimension. In fact, between the XII and the XV dimensional Plane of Existence one finds oneself in the IV spatial dimension (a place outside of three-dimensional space and in which it is still possible to create materially) and temporal dimension (the perception of time for those who are consciously aware of the dimensional Planes is different from those who are not aware of them). In those Planes of Existence we therefore find ourselves in a sort of transition state. Once we cross the membrane of the XVI dimension of Existence, we find ourselves in the V dimension.

In the subsequent planes, up to the XX, one can learn total multidimensionality, i.e. circular time and the non-place.

The XVII dimensional Plane of Existence is the World of Dimensions. We enter this Plane as though through an illuminated corridor and, once we reach it, the light is white, luminous, but not blinding. Here there is no-one else except the awareness of ourselves but, however, we sense the presence of an infinite number of Beings with whom we can interact. Here it is given us to interact with Beings of all dimensions, both our own, and of other Universes. Here we find the knowledge and study of the technologies of the infinite Universes, just as in the VI Plane of Existence we find the laws of one's own Universe.

All technology, when it is extremely advanced, is Spiritual and no longer material. At times we can sense the presence of equipment made of light and that emanates flashes of light. Here infinite healing techniques can be experienced, all aimed at achieving maximum balance. Only those who have reached the turning point in their evolution will be allowed to consciously access this Plane, after having crossed the threshold of multidimensionality.

The XVIII dimensional Plane of Existence allows us to understand Infinite Beauty. The understanding, love and creation of Beauty. Beauty is the infinite projection of Infinite. Here it is possible to comprehend the Beauty of our Life Goal, so that it is created with and in Beauty right from the start.

With the XIX Plane of Existence one accesses the Infinite Quantum Possibilities. In the infinite white light, one is granted the sight, in a fraction of a second, of Beings who then immediately disappear. Each of these Beings is one of the visitor's selves. They are the visitor himself in one of his infinite quantum possibilities. Once we have had access to this dimensional Plane, we can ask to see the self of the quantum possibility that we want to live and therefore speak to it requesting to vibrate with its own intensity. Or, more precisely, to resonate with him - or her if they present themselves in female form - if that is how we want to be in our new life.

The XX dimensional Plane of Existence is the Plane in which we can Hear the Infinite. In it, the Infinite speaks and grants the possibility of reading everything and everyone on whichever quantum line they may be and onto whichever quantum line they move. From here it is possible to very accurately read the quantum lines of linear time, which are often fleeting. Since here we are in the Infinite, we can know everything that has happened or that will happen, everywhere and for anyone, with extreme accuracy.

Once the journey has been made consciously, in all the dimensional Planes described and in which the human being is capable of existing, we realise that as we gradually approach the perception of Infinite, the brain also reads the body that travels together with the mind. In fact, from the fourth dimensional Plane of Existence up to the XV, we travel especially with the perception of our Essence, whereas from the XVI Plane of Existence onwards we also perceive the presence of the body, because the brain has - at that point - learnt (and the cells understood), that the body is capable of travelling together with the entire Being. Moreover, as we consciously enter all the dimensional planes, the concept of divine as being unitary, consisting of a single supreme Essence, becomes more and more remote. The human cells feel that everything is pervaded by the All; there is the essence of something everywhere and there is no end to this. The concept of an arrival point is no more; the desire to know "the vertex" has drifted away. We clearly perceive that in the Infinite, everything is Infinite. The anxiety of moving higher and higher subsides. The new path to entering Infinity begins.

Entering the Infinite in finite form, i.e. with the body.

This is a great change, and regards the people of this historical period of time.

5.

The Personal Vibration.

This is like a weave that lies all around and within all living beings and that derives from the electromagnetic field produced by every single person while emanating their own vital energy.

The Personal Vibration interfaces with the much vaster electromagnetic field produced together by all things, people, elements, etc., that we simply call our Universe, integrating with what I call the Total Emanation, expression of the Universe itself.

In order to describe the way the human brain visualises the Personal Vibration, imagine a globe in which the meridians and the parallels make up a thick and close-knit mesh that creates the globe itself. The spherical shape that is thus created surrounds the being that emanates it, within the radius of approximately two metres from their physical body, which is enclosed yet free to move within it.

At each point at which these meridians and parallels intersect there is of the being that emanates them and every being is both in the globe and is the globe itself. In other words, this close-knit mesh that surrounds us is simply called the time-space dimension.

This means that the meridians and the parallels that make up the Personal Vibration consist of two elements: space and time, and that the luminous points that are created at the intersections of these elements are none other than cross-points of space and time.

As we said before, everything that concerns human beings belongs to a specific level of a certain dimensional Plane of Existence. For example money, jobs, sentimental relationships...all belong to a specific level, and each of them falls at a certain point of the space-time intersection of the Personal Vibration that envelopes the person.

Because we said that there is something that concerns us at each point of intersection, it is easy to comprehend that the luminous points are the parts of us that are always at the right place at the right time!

That is fantastic, because it once again confirms that each of us is the maker of our reality, in that we are all always in exactly the best place and time for us, this meaning that if we remove the conventions of space and time, we are always in the "here and now".

Therefore, even instantly creating our own reality is a prerogative that belongs to each and every one of us. In fact, once the Personal Vibration has been visualised, it is enough to see ourselves, by means of the delta waves, projecting ourselves with all of our being, our body included, into that particular point of light of the Personal Vibration, creating what there is of ourselves in that particular point.

To make a practical example, if someone wants to create the best possible sentimental relationship there is for them, they can go to their own Vibration - by consciously using the delta waves - and in it search for the luminous point that contains their relationship and the way in which it best suits their needs. The person will then fully and consciously acquire the image of what the best sentimental relationship is for them, according to their own personal structure and their own personal needs. Once this has been done, the next step will consist of projecting this image into the area of their front lobes and bringing it into their immediate reality. This involves projecting it into their own space, thus attracting the best person for them, who allows them to fulfil the kind of sentimental relationship found in their Vibration.

It is essential to have the images of things within us; without them, in fact, our brain would not know what to look for. It is a fact that the image of all things is contained in everybody's Personal Vibration, even if it is not present in that person's brain. This phenomenon is due to the holographic nature of the Universe, by which Everything is in everything, therefore all images were once present in the human brain and if, as a result of a particular dissonance, it has lost some of them, they were not lost entirely, but instead can be found in certain space-time intersection points of the person's Personal Vibration.

Therefore once we know the image of something, of a feeling, of an emotion, of a symbol...we can create it in our own reality, because we retain the memory of it. This means that once the missing image has been found, this reactivates the cellular memory already contained in the DNA, making it real. In other words, it reawakens something that exists, but that was lost, in meaning or in use, by the brain.

From this point of view, the words in the ancient texts that bear the knowledge by which God created Man in His Likeness can be interpreted in the following way: the Universe created the human being according to the image that It had, in other words according to the image that was present in the Total Emanation. This implies that, on the basis of the holographic principle of the Universe according to which human beings are little universes in their turn, they are therefore capable of co-creating reality, once they have the image of it. Co-creators, therefore, because the memory of the act of creation is has always been present in everyone's cells, because we ourselves were born as a result of an act of creation that the Universe performed.

In brief, we can say that we are creators because we were created.

What we have said until now regarding one's Personal Vibration applies especially in the "old grid", but the ways of acquiring the image and of projecting it to the brain can equally be used in the "new grid". In this grid in fact the Personal Vibration continues to exist, even if it is slightly different and is visualised by the human brain like a looser mesh made of extremely fine threads of photons, and through which the quantum leap can be made...

In actual fact, the Personal Emanation does not change; what does change is how we perceive it once our dissonances have been resolved.

6. The Universal Vibration.

The Universal Vibration is the sum of the Emanations of all the beings of the Universe. One uses it in order to succeed in opening what I call space-time Star Gates. The opening of these "gates" allows one to access the knowledge of one's sense-project and Life Goal within the Great Plan of the Universe.

Just like the Personal Vibration, the Universal Vibration is also visualised by the human brain like a sphere made up of intersecting horizontal and vertical circles, made up in their turn of extremely thin photon quanta which lend these "meridians" and "parallels" the characteristic of iridescent luminosity. The photons that create the meridians are the electromagnetic vibration of light itself; the photons that create the parallels are the energy produced by the dimensional Planes of Existence, whereas the intersection points between the Planes and Vibrations are called "dimensional points of intersection".

The images of everything that exists in the Universe are contained in the dimensional points of intersection, including forms that have never been in the human brain, such as the image of Infinite or other similar things. When we say that some images have never been in the human brain, what we are actually saying is that they have not been there for several thousands of years, i.e. for so long that they have actually been excluded even from the Personal Vibration of human beings, as well as from their brains.

Through the conscious use of the Universal Vibration, it is therefore possible to project images into the brain, so as to allow the expansion of awareness beyond the limits established up until now by the conventions of space and time, which carry with them the convention of finite.

The expansion of awareness is very important for all human beings, because in that way we are able to access and accelerate the "passage to the Fifth Dimension".

7. Images create life.

In the beginning was the Word.

This is the beginning of the All. This phrase appears in the most important ancient western and oriental texts. In various forms, in parables or circumlocutions, what the sacred texts have always said is that reality is created through words. Each time we use words, we create our reality.

In fact, the brain associates the words that are pronounced with images that it knows and that it immediately begins to make real.

Life, as it appears, is an emanation of the Word, and the reflection of the Word on reality is image.

Therefore, a person's life depends on the image that he has of it, and on the words he uses to define it.

It thus follows that if he wants to change his life, or parts of it, all he has to do is change the image he has of it, replacing it with the image of life he desires. One therefore needs to know how to create the "new" image. In this connection, it is important to emphasize once again that no image is ever new to anyone, because the brain contains the form of everything there is in the Universe; this is due to holographicity. Therefore Everything is in everything. Generally, however, one is only aware of a part of the images contained in the brain, and this is because one only uses a part that, on average, corresponds to 5% of its full potential.

When we arrive at consciously using increasingly substantial parts of the brain, we have access to what has been called the Personal Vibration, which contains everything about ourselves in the form of images.

We said that having access to the Personal Vibration means having access to the electromagnetic field that emanates from all beings and that is the vibration of the matter itself of which they are composed or, more precisely, the vibration of the chemical and nuclear bonds that hold the molecules together of which, for example, the human body is composed.

We said that, by interacting with the Vibration and going into it, we can choose the image we consciously want to emit into our brain, so that reality begins to resemble the image we have chosen.

We also said that the image of the All is contained in the Vibration, and that each time the brain recognises the image of something, it begins to create it.

We can therefore say that in order to change our reality, it is enough to use images that differ with respect to those that have been used up until now. Being able to do so, however, calls for two basic choices:

1) the decision to use the greatest possible percentage of the potential of our brain, and

2) the choice to look for and find the image of Beauty in the All.

The second choice especially will change people's lives. Since beauty is in everything and if we are capable of recognising it, we can tap into the Infinite Energy of the Universe, thus raising our vibration.

The image of beauty.

Regardless of any cultural, stylistic and cognitive metaphor, the delta evolution path places particular importance on the beauty and youth of the body, in other words, on our physical form, this being our means of existence in what we call our world.

We said earlier that Beauty is in Everything, and that if human beings are capable of recognising it, we are capable of tapping into the Infinite Energy of the Universe and, consequently, raising our vibrations.

The concept of Beauty as a form of the divine or -if expressed in terms of energy- as a form of maximum vibratory intensity, is present in the biological memory of all human beings. In fact, in all the ancient history of humanity, Beauty has always been recognised as the intermediary between human beings and the divine. When we think of the classical Greek myth, Helen, who represents the archetype of beauty in the expression of its immense energy both to create and to destroy, was a demi-goddess daughter of Zeus and of a mortal, standing to indicate that when we have beauty, see beauty or simply sense beauty, we already are on the divine plane.

Beauty is what draws man to God, therefore, if we are able to glimpse it, we create harmony and balance.

Once we reach a certain degree of evolution, beauty becomes an impulse of the soul; it is illumination that is reflected in the body. It is the awareness that materialises, becoming visible to the human eye through the beauty of the physical body. There can be no illumination, awareness, spiritual evolution, without it resulting in physical beauty, because only thus can we have access to the Unity, we are One with the Universe, in fact, only thus, is the Being complete, having overcome dualism and contrast.

Only thus, in other words, are we in harmony with ourselves. Once we are able to find Beauty within, and bring the image to the Brain in a conscious manner, we will have access to the energy of the All, and our vibrations will be raised. This means that the electromagnetic energy that our body emanates will vibrate at a higher vibration and the Personal Vibration will become consciously accessible in increasingly greater parts.

What we have said and repeated until now is that if we consciously take different images to the brain, we can change our life. If this is true, we can also say that if we take the image of beauty of the physical body to our brain, it will start to raise the vibrations of all our being, because as a result of the holographic principle and the principle of oneness, the brain will perceive the beauty of the physical body as the beauty of the Universe.

And so it is.

The concept of beauty is the leitmotiv of this whole research.

Beauty is what draws man to his divine part, because if he is capable of glimpsing it, he is also capable of creating harmony and balance. The most important thing is that beauty is in each one of us, who is as they are, therefore a perfect being because they are such.

Each of us different, each of us unique and unrepeatable.

CHAPTER IV

THE BRAIN IN THE ANCIENT WEAVE

In order to go beyond something, anything, we must first fully accept what that thing is, in all of its forms and manifestations.

And so, in order to live between Worlds, we must first and foremost be ready to live in the World. Accept its rules and facets, up to their smallest contradiction, after having understood them. Only like this can we understand its essence and nature, setting out on our path in order to go beyond.

We are here, now and we must live whatever presents itself to us, to the fullest. We accepted it when we were born and we must accept it always, while we remain here. We cannot work well for something that we do not accept or understand, and so first we must accept our human and earthy reality and live it all the way. We must profoundly understand the task the Universe has entrusted to us here and now, and only afterwards can we continue the path. If we do not do this first, everything that follows will just be a desperate attempt to flee from the present. Each time we try and escape our reality, we die at some level, be it intellectual, spiritual, emotional or corporeal. Death is only the easiest way we have to distance ourselves from what we do not accept, first and foremost from ourselves.

There is an intermediate state before one passes to death from life, and that state is called malaise. When we talk of malaise in this book, we do not refer to physical disease, but to the state of disharmony, at any level, that creates problems in people's lives. Malaise is the lack of harmony. We must be in tune and in harmony with our Universe, because the lack of harmony, at whatever level it may be, corresponds to malaise.

Healing is being attuned to the Universe, being in harmony with the All, being aware of ourselves. Healing is the beginning of the Path, and in order to walk it we must know ourselves profoundly.

To facilitate this first and important step in personal evolution, the following chapter will introduce some of the issues that concern the biological behaviour of the human brain and the individual ways used to respond to outside influences, thus providing the reader with the first, useful tools.

1. Biological functioning of the brain.

Because the first thing we need to so is to bring the automatic responses contained in our brain to a conscious level, those that were inherited biologically or learnt through behavioural models -in any case coming from outside sources-, it is important to begin by schematically explaining how the human brain functions biologically and the automatism of the responses that a person gives to external stimuli.

It is important to emphasize that although we are discussing the biological behaviour of the human brain as we know it now, the goal of the Law of Delta is to provide us with answers that will lead the brain to developing capabilities unknown or, more precisely, inactive until now, that will lead to a different kind of behaviour of the brain itself, which will no longer respond to the biological laws as we know them now.

However, as we said earlier, in order to evolve we must profoundly understand the starting point, and particularly the biological evolution of the human brain, as described for the purpose hereunder.

The brainstem, also known as the "R-complex", seems to have been the first to appear in evolution; it is in fact the brain that evolved in all animal species, with no exception.

It is also called the automatic brain, because it does not think.

The information contained in the oldest part of the brain ensures the human being's survival of the human race. This also applies to all animals since they all, indiscriminately, have a brainstem. The information it contains corresponds to the fundamental principles of life on the corporeal-material level, which still apply today. These can be summarised in three broad groups in which the dynamics unfold regarding the relationship that exists in nature between predator and prey, between reproduction and sexuality and between nutrition and food.

Any problem that human beings have, at any level, can biologically be traced to one or more groups of the dynamics described above. This first primitive brain, programmed simply to eat, defend from attack, breath, reproduce, has allowed the human race to survive. It is said that it does not think and has been called automatic; in fact it has a sensor, a kind of alarm that triggers all survival reactions. This sensor is, precisely, automatic. It is indispensably automatic, because this part of the brain contains base reactions that must be activated instantaneously, in order to ensure survival itself.

The Midbrain: this is a more evolved brain, which appears in evolution subsequently. It is present in all animals except reptiles, and its main task is that of managing emotions. In evolution, the appearance of this brain distinguishes the passage to a subsequent phase. When we imagine animals that live in a herd, for example, we can understand how the evolution of this part of the brain distinguishes the passage from a previous individual situation, in the almost total absence of inter-relations, to a situation in which a scale of values and relations -a herd or flock- is created. But the real characteristic of this part of the human brain is the onset of emotion.

At this point of evolution, this brain provides access to emotion and one acquires the instinct to flee. Moreover, there is a zone of this brain that manages the automatic responses learnt during the course of human evolution and that are contained in and transmitted from this zone. It is a sort of archive in which all the responses learnt by the human being himself and by others of his kind before them are stored. These are responses that aim at resolving all contingent situations and all emotional states. They are called automatic, because the brain is capable of providing them instantaneously, without having to pass through reason. The midbrain, thanks to its memories, as well as transmitting simple needs, is also capable of transmitting emotion or, more precisely, the desire related to that emotion. In fact, as we said earlier, whereas in the brainstem a need is reduced to "nutrition", in the midbrain the need is processed and interfaced with emotions and then expressed in the form of a desire for a specific type of food. Beliefs are also held in the midbrain, the things we learn from the outside.

The zone of the brain called the *cerebral cortex* seems to have come last in order of time in the evolution of human beings. This part is personal and all the commands in it correspond to relational problems and therefore to concepts of individuality. It therefore represents the most recent part of the human brain and is the seat of thought and of that which is commonly called rationality. In fact, the ability to think came last of all, once the human being had perfected the ability to foresee.

At a certain point in evolution, the man-hunter could no longer have been without thought, in that he was not strong and fast enough with respect to the animals he hunted, and therefore the best way he had to compensate for these weaknesses in his physical structure was that of anticipating the movements and behaviour of his prey. Viewed in this light, one can initially say that thought resembles a kind of supplementary sense that enabled human beings to survive and that only subsequently developed in the refined and Machiavellian manner as we know it now, thanks to the development of the cortex itself and of the front lobes.

2. Needs.

O Now that we have a general idea of the structure of the organ that is the prime subject of this study, we can start out on our path of awareness. In order to travel the path towards knowing ourselves, we must first be aware of what our real needs are. It is therefore indispensable to know what needs are part of ourselves and of our structure and uniqueness. Once fulfilled, these are the needs that provide total and perfect harmony with ourselves.

Way too often people do not satisfy these needs, because they do not correspond to what we have learnt, or to what we have been led to believe are everybody's needs. They do not correspond to what we are told by advertising, culture, tradition, the social environment in which we live, the convictions, the biological memories we inherit, the emotional sequences left incomplete by our parents ... People therefore do not satisfy their needs as a result of their beliefs, of the things they learn from the family or from society, or that they have inherited from their parents and they from theirs...

Everyone is different and everyone has different needs, not everyone has the same priorities and this means we must know ourselves and our needs before we can start creating our reality.

Even if we have understood is possible for everyone, that there is no sense to the inherited way of thinking by which we have to suffer in the world, that we can have everything we want...we must still be careful to avoid continuing to make the mistake of thinking that what we want or need is the same for everyone. Because, to the contrary, this would be the same mistake that has been repeated until now and that has only led to unhappiness: that of thinking we are all the same.

Creating our own reality means satisfying our needs, because these are the ones that generate desires. Because if each of us looks deep inside ourselves, we will discover that not everyone needs to possess a yacht or a luxury car; there are people who have different needs. If we reflect on that, we can understand why easy recipes for achieving happiness, by which everyone can become a millionaire and by which everyone once again lives the same life as everyone else, cannot work. They cannot work, simply because this would mean repeating the same scheme of standardisation. At another level, but again the same scheme.

Thus, most people feel frustrated about not being able to become millionaires and they do not realise that this does not happen only because they do not really desire it. In other words, they do not feel such a strong need as to transform it into desire, but are simply expressing an idea in the attempt to conform to others. It is wise to always remember that when we try and conform our needs to those of others, we are no longer fulfilling our true needs and we therefore automatically enter into dissonance.

Dissonance means losing harmony with ourselves and with the Universe. Dissonance and therefore the lack of harmony lead to malaise.

This means that external influence, but especially the value that each individual person gives to external influence, can condition their wellbeing or their malaise. To better explain this seemingly abstruse concept, a number of scientific discoveries on the behaviour of the cells in the human body must be introduced.

Approximately fifteen years ago, researchers in the field of Biology discovered that when the nucleus was removed from human cells, they continued to live anyway. It became clear, for the first time, that the brain of cells is not located in the nucleus as had been thought until then, but instead is situated in the membrane. Since every human being is made up of fifty thousand million cells, this discovery bears substantial importance in understanding the laws that govern the exchange of cellular information and the way life is conducted, from a biological point of view.

Understanding all this becomes even clearer when put into relation with the discoveries made in psychobiology over the past few decades. This discipline has established that not only do we inherit so-called genetic characteristics from our forefathers, such as the colour of their eyes or their hair...but we also inherit their so-called genetic memories, in other words their habits, dissonances, automatic responses, harmonies and disharmonies...

All the information inherited is, naturally, found within our cells, and these are coordinated by our brain, which is composed of cells. Therefore every time we say that a disease is hereditary, what we are saying is that the ancient memory of the dissonance that led to the disharmony and, therefore, to that disease, exists in the cells. Seen from this point of view, every illness can be considered hereditary, because memory of the dissonance can also be found in the previous generations, despite it not presenting itself in the recognised form and with the same symptoms.

The brain and the human body, meaning every single one of its cells, (always remember the holographic principle of the Universe) can, therefore, be described as a large computer that has an enormous file, some parts of which have been turning for millions of years, other parts for thousands of years, others for centuries or generations, and yet others for the past twenty-thirty-forty years, according to how old we are.

This file holds all the information that our cells have accumulated over time from the experience of those who preceded us; this information has been useful, some still is useful, other parts of it have become superfluous or harmful. However, it continues to be there and to function automatically, even if the person does not want it or believe he does not want it.   Therefore, the surrounding environment has influenced man us from the beginning of time, because the responses processed by his forefathers or by the person himself were aimed at improving life within the habitat around him. The fact that the brain of cells is situated in the membrane simply helps us to understand how the information from the outside, -from the surrounding environment-passed into the person and his predecessors, simply through "feeling". This information then remained in the cells themselves and was passed down from generation to generation; since we said that cells are like big computers, other information continues to be saved from scratch by each generation, it being absorbed from the outside environment. Once we understand this mechanism and become aware of it, we are already *en route* to harmony or, if you rather, to healing, first of all because we understand that we have been walking along a sole line of quantum possibilities for thousands of years and, secondly, which possibility had been followed until now. Moreover, it is easy to see that, just like we do with a computer, we can change the files that are damaged or that simply hinder our current life. There is no need to change the entire program. In fact, some of the automatic responses, inherited or acquired, are still useful, perhaps they save our lives, and therefore it is good to keep them, at least until all the potential innate in the brain has been acquired.

To initiate change, it is therefore necessary to identify our needs, which are personal and combined differently from person to person. This will enable us to understand which of the files it is good for each one to keep and which, on the other hand, have become obsolete; which have been used correctly and which however have been used incorrectly.

There are various ways of identifying what people's real needs are: by observing the person and their behaviour, as well as by analysing their life and habits. By means of the Law of Delta, we acquire the tools required to fully understand who the person within really is and what their dissonances really are. Once our personal needs have been identified, the next step will be that of identifying what the active base dissonance is.

Dissonances.

3.

People have what seem to be an infinite number of dissonances of various kinds, but in actual fact they can all be reduced to three broad base categories, which represent humanity's fears.

Dissonances, in fact, are fears.

On a bodily and material level, fears give rise to chemical exchanges which, in a way, saturate cells in such a way that they feel as though they are sufficiently nourished and satiated and do not feel the need to change the situation.

On a chemical level, fears behave in such a way that enough space for anything else remains in the cells not and so, paradoxically, those who have the most fear (or dissonances, if we prefer), are the least inclined to free themselves of them and change life, despite their being convinced that the change will be for the better. In the midst of the great amount of dissonances, it is therefore important to succeed in identifying the base conflict, in other words the triggering fear. This makes it possible to simplify and accelerate our self-awareness process and free ourselves of that sole fear on which all the others depend. All dissonances can thus be abandoned simultaneously, instead of resolving one dissonance at a time without realising they are all really different facets of the same fear.

The three base fears and the respective dissonances connected to them are illustrated below:

The fear of abandonment is an ancestral fear, of an animal kind. It is the biological fear that the animal cub feels in being abandoned by its mother or the herd. All human beings inherit the biological animal memory by which the brain knows that if they are abandoned by the mother or the herd, the cub will fall victim to predators and will therefore die. The part of the brain in which this memory is found is the single oldest part of the human brain, i.e. the brainstem. Where this fear lies indicates that it is the single greatest and oldest fear that human beings have, because it is contained in their biological heritage and has therefore in their memory for thousands of years, in other words, since when they were still animals. All disharmonies of abandonment derive from this fear, whether they concern being abandoned or abandoning. They present themselves in various ways, including aggressiveness, obesity, bulimia and combativeness. Subservience and anorexia are also part of the dissonances triggered by this fear.

The second fear is a more recent biological memory than the previous one, and appears when the person begins to relate with the group. He measures himself with others, beginning to doubt and wondering whether he is up to those around him. This fear implies a judgment of value and all devaluation disharmonies arise from it, particularly sexual ones and ones concerning judgment, sacrifice and the annulment of the emotional part within. People who are very much in the masculine and who have almost entirely abandoned their feminine side fall under this kind of fear. Moreover, every time one identifies this kind of fear, it is easy to find that the person afflicted has ties connected to vows or oaths made in the past, even by their forefathers. Therefore people who have this second fear often find ancestors in their lineage with profound connections to spirituality or who were attached to strong ideals, from which the fear of not measuring up to them derives.

The third is the fear of surrender and letting go. This is the most recent fear in biological memory, dating back to this era in time and that is the epitome of all the disharmonies that manifest through what is, normally, called depression, at whatever degree of intensity it has reached.

Generally speaking, one finds a single basic fear in each person; it sometimes happens to find more than one, but it is always possible to identify the stronger and therefore main one, whereas the others are secondary or derive from the endurance of the dissonance triggered by the first.

4. Automatic responses.

Having observed the biological working of the brain and its three parts and having discussed dissonance, we must now go into greater detail as to what causes it in biological terms. Once we understand the mechanism of automatic responses, it will be easier to grasp the above, as well as the logic of the parts discussed until now.

When, upon the onset of a need in the brainstem, a corresponding emotion and then thought follow that make it possible to realise and then fulfil that need, there will be no problem for the person. In that case, we can say that the brain found the same information in every one of its parts (the brainstem, the midbrain and the cortex) and that the "three brains" vibrate with the same vibration. If, however, the information contained in the three areas of the brain is unalike, they would vibrate at different frequencies. A discrepancy of this kind would cause a problem in the person, and therefore the onset of dissonance.

In fact, by immediately sensing the anomaly in the vibrations, the brainstem sends an alarm to the midbrain which then activates the automatic response. Within it, the midbrain has something that can be imagined like an archive of all the solutions processed by those who have lived before us in time. Since, as we said, emotions are generated in the midbrain, the solutions it contains concern emotional sequences experienced by the person's parents, forefathers, communities, clans, and so on. These filed solutions are called "automatic responses", because they are emitted by the brain in a fraction of a second, in order to resolve the problem created by the non-alignment of its three parts. Technically speaking, in this way the brain finds its own solution, but in actual fact, as a result of this mechanism, another kind of problem arises due to the fact that the response that comes from the learnt, or inherited, biological memories is yes, the good response, but to an old problem, in other words the good response transposed from the old problem onto the current one. The response is therefore technically perfect with respect to the survival of the species, but is entirely unsuited for ensuring a pleasant life. Moreover, this mechanism is such that behavioural patterns are repeated endlessly. It is therefore as clear as it is important, indeed indispensable, that each time an automatic response is triggered, we are aware of it. Like that, in fact, nothing prevents us from giving a new response, one of our own and one that is better suited for our needs for a pleasant life.

To do so, however, we must first be aware that we are giving an automatic response, which by definition, is unaware.

By means of the delta waves, it is possible to bring the inherited automatic responses to our awareness, so that it becomes perfectly clear to us which are still appropriate for our current life and which, to the contrary, should be changed.

5. Bringing all the automatic responses to our awareness.

By using the Law of Delta, the image for obtaining the immediate awareness of one's automatic responses can be brought to the brain. This image is enough to change the situation. Because the human brain takes in and recognises images, it will immediately begin to resemble the projected image, and the sign that all this is happening will be that of a sudden change in our degree of self-awareness. We find ourselves better and immediately understanding our behaviour and our ways of interacting with our reality.

We will initially be aware of the automatic response and its inappropriateness immediately after giving it, then gradually we will become aware of it just before we give it. The fraction of time in which we are aware will be the exact moment in which the new response will be inserted, the one which is better suited to us and always immediate and which, in its turn, will become automatic.

6. Awareness and automatic responses.

The phase during which we become aware of our automatic responses is very important, because it is then that we have the chance of observing our old schemes and how most of them prevent us from generating a happy, joyous, adventurous life... We must first however turn our attention towards distinguishing the emotions or reactions that are still useful in our life from those that are of no use or that are harmful to us. In fact, a fit of anger at the right time is always useful, in that it "saves" us from a situation that is dangerous to us. Often the danger lies simply in the frustration or lack of authenticity that arises from our behaving in such a way as to suppress this anger, and therefore in such a way as to repress emotions judged to be "bad".

The act of bringing automatic responses to our awareness aims particularly at releasing the emotions, present in everyone, allowing them to be brought to light without control, showing themselves in all their intensity, contrast, diversity, depth, etc.

Emotions are simply emotions, and there are no good emotions or bad emotions: they can only be evaluated for their efficacy in helping us live our lives well. They are effective when they are used or, better, when they are allowed to flow naturally without subjecting them to the control of the rational mind. If left free, emotions interact with reality, creating situations of the utmost good for the person feeling them. To make a practical example, think of anger: if it is judged and considered "bad", we will do everything possible to control it and push it deep down into ourselves, without realising that it is there in any case and that it continues to work from within, until it reaches a point that it permeates our whole life. Thus, we will end up with a person whose every behaviour clearly indicates they are trying to hold their anger back; as time goes by, we will have a person who is on the brim of having their anger explode and finally, we will have someone whose gestures are particularly ferocious, because their anger has exploded.

In these cases, people will it was an act of lunacy o something similar by someone particularly good and calm. In actual fact, this deed will have been done by someone who, day after day, unconsciously cultivated their anger until it exploded. When we stop controlling our emotions and let them out as they are and at the time we feel them, they will not do any harm at all, because they will not be carrying a particularly heavy load but, to the contrary, will greatly ease people's lives.

During this phase, we can understand our behavioural schemes and our relationships with our emotions and bodies, because they will be free of control.

7. Changing automatic programming.

Once we have understood the behavioural mechanisms that keep us in perpetual conflict, we can free ourselves of them forever. Automatic responses were inserted into the human brain in order to ensure biological survival, and even if most of these are no longer of any use, it is still useful for people to eat, sleep, defend themselves...automatically. Therefore, only a few of the responses will need to be changed, whereas most of them continue – at least for a while – to be part of the cognitive baggage of human beings.

Once changed, the old automatic responses do not immediately disappear from life's horizon, but remain for a little while longer. Terminating their task in the Universe, they complete the person's emotional release. In fact, in those who have lived in dissonance, the emotional part has been bridled and kept semi-dormant for a long time. For approximately one month after the automatic memories that had become futile are removed, the brain feels emptied of everything superfluous that was in it: fears, dissonances, unfinished and inherited emotional sequences, beliefs and convictions that hinder the quality of our current life.

We said that fears activate chemical exchanges in the cells, such as to saturate them; in this situation therefore the cells will be emptied of this saturation and will find themselves in the best state for receiving and transmitting new information. They will therefore be able to recognise the conscious programs for the best instantaneous solution to the true needs of the person to whom they belong. With the new programs made automatic, it is easy for us to change our way of relating to situations that previously led to dissonance. We have increased contact with the unknown parts of ourselves and experience a peacefulness that emanates from the deepest layers of our Being. Having lived without that feeling for such a long time now, this peacefulness is at times so intense as to even initially feel strange.

This is the beginning of total self-awareness.

8. Strategies for using Energy

After having outlined the biological ways of behaviour of human beings in the previous paragraphs, another typical kind of behaviour will be illustrated below, which takes place by means of energy exchanges between people.

There is an infinite source of Energy which is the Universe itself, but man has forgotten how to make unconditional use of it and has therefore invented strategies to steal it from others, convinced that this is the only way of obtaining it. In daily life, this mechanism concretises in a continuous alternation of moments of extreme tiredness with moments of intense activity, according to whether energy was taken or given.

The four main strategies in use by human beings during this historic period of time, which have the purpose of taking energy from others, are described below. At the end of the description, it will be easy for the reader to identify their own way, put into place after birth and therefore learnt within their own family unit. The best solution is to learn to remember the absolutely best way for everyone, which is that of taking energy from the inexhaustible source of the Universe. Learning to receive energy in the right way is very important, because by doing so one is always capable of giving energy without ever lacking in it and, on a cerebral level, this is the same as being able to give, and to receive, emotions. Because we said that emotion is what creates reality through desire, learning to exchange energy in the best possible way for everyone is the same thing as learning to create our reality in relation to our surroundings.

As we said, therefore, four main types of energy exchanges take place within groups of people, and these correspond to the basic behavioural schemes to which most human behaviour can be traced when involved in interpersonal relations. There will be active-aggressive people or passive people, according to the types of strategy they adopt in their daily lives:

The first category is that of Judicators: these are people who are naturally active-aggressive, who continuously ask questions, and who intrude in the life of others so as then to judge the behaviour and actions of their "victims". This is the way in which they steal energy.

Intimidators: these are also active-aggressive

people, who tend to threaten both using words and the tone of their voice, as well as through real acts of physical intimidation through which they take energy from others, thus embezzling it.

Victims: these belong to the group of so-called passive people, who continuously tell others everything bad or negative in their opinion that happens in their life. They are capable of making the person listening to them feel responsible and impotent if they do not find a way of helping them. That is how they take other people's energy, i.e. either by having others help them or by making them feel guilty.

The Taciturns hook people by their secretiveness, arousing curiosity in others so that they go to a great extent to understand what it is that torments such an introverted person, thus surrendering their energy to them.

In a group, the four types interact with each other, each creating their complementary figure, in fact:

People who are too reserved create judicators;

Judicators create people who are too reserved;

Intimidators create victims;

Victims create intimidators.

The above has implications of outstanding reach. The word "create" in fact, is used not only with the literal meaning (that it assumes when, for example, children in a family group model themselves in a complementary manner with respect to their parents), but also in the sense of attracting, when we refer to other people. As we said, this has very important implications that should make us reflect on the fact that everyone creates their reality, attracting complementary people into their life. It is very important therefore to stop using the energy strategy once and for all, especially for those who use the victim strategy, because inevitably sooner or later someone from the aggressive categories will arrive to take advantage of them in every way possible.

The two examples below are given to indicate a possible solution for freeing ourselves from the habit of exploiting other people's energy, and to provide a solution for tapping into the energy of the Universe. Try it as a game, and then verify the results!

The way to stop ourselves from exploiting other people's energy is that of identifying our usual method and consciously taking care not to activate it, whereas in order for other people not to take our energy away, it is enough to simply recognise the group they belong to and call them by that name. This can also be done by including it in the middle of a playful sentence, such as "I see you feel more of a victim than usual today", or "Since you enjoy being so secretive, I'll leave you in peace". By means of sentences like this, the automatic response mechanism that we mentioned earlier is put to beneficial use, because playful sentences such as these put the interlocutor in their automatic response mode, leading them to believe they have to answer, and this distracts their brain entirely from the energy strategy mode. Because the energy strategy is also an automatic response, these sentences trigger an automatic response that is even more deeply-rooted and ingrained in the person, because it belongs to the innate biological memories, whereas the strategy has been learnt. A mechanism is thus activated which leads to the result of freeing ourselves, at least temporarily, from the person who wants to take away our energy.

Freeing ourselves definitively from our own strategy and from that of other's people's is the correct objective to be pursued, as can be seen in the following paragraphs.

9. Active energy strategies.

First of all, your way of taking energy from others, i.e. your energy strategy, needs to be identified. To do this, simply assimilate the concept and the four methods described above and then ask yourself questions like: "In all honesty, what can I say is my way of taking energy? Which of these four is my usual way of interacting with others?"

Close observation will quickly highlight the kind of strategy you use. If it is difficult for you to identify it or you are undecided between two methods, you will in any case be able to immediately pinpoint which of them you recur to most often, and refer to this. The beauty of these strategies is, in fact, that once they have been recognised they become extremely easy to identify and therefore also to abandon.

10. The energy strategy and the infinite demand.

The way in which energy is exchanged by taking it from others, i.e. the type of strategy used to take energy, corresponds to the infinite demand everybody develops in their childhood, but viewed from a different dimensional Plane of Existence. In fact, whereas the infinite demand conditions the material, sexual and emotional plane, the energy strategy conditions the emotional, intellectual and spiritual plane and, in relation to behaviour, interacts precisely at an energetic level. Before we resume the explanation of this concept, we need to discuss the infinite demand in greater detail and see what this term really means.

Infinite demand: this arises in us as a result of what we feel we did not receive when we were infants and that we clearly never will be able to have, but that we continuously ask everyone for. We will never be able to have it because it concerns something we felt as a baby, which obviously we no longer are.

Generally, the most common infinite demands are one of two kinds: being recognised or being accepted.

Recognition: it is the father who gives recognition. If he recognises his child, the child will possess the image of "male" and of everything that is masculine; if he does not recognise his child – even by simply thinking for an instant, when the child was conceived or just as he was born, that he did not want him or did not want him as he was – the person will not have the image of "male" in his brain, nor of anything connected with being masculine. Therefore, if it is a girl who is not recognised, she will use (once she becomes an adult without ever having possessed the image of male), portrayals of models from the current-day culture, and will look for an ideal man who does not exist.

If, on the other hand, the child is a boy, when he grows up he will be a man who tries to incarnate the man that his woman has in mind, and will therefore be just as unreal as the image of "male" that the woman herself has.

Acceptance: this concerns the mother; it is she who must accept the child as he is. If she accepts him, the child will possess the image of "female" and of everything that concerns being feminine. If the child did not feel accepted, or if he really was not accepted (remember that to the brain there is no difference between feeling and being), all the images concerning "female" will be lacking in the brain.

If the unaccepted child was a boy, as a man he will look for a woman who corresponds to cultural, television models.

If, on the other hand, the child was a girl, she will be a woman who tries to impersonate the feminine ideal that her men have in mind, and will thus expect the man to define her as a woman.

A lesser amount of people were neither accepted nor recognised, and these therefore do not possess either the image of "male" or that of "female".

In all cases of infinite demand, one of the two following modes is always activated each time we interact with others and even with ourselves:

either we apply the strategy and then tell lies about our identity, for example taking images borrowed from the outside,

or we renounce acceptance and/or recognition, in which case we are sacrificing ourselves.

In both cases, there remains a part of us that is not confessed to the other person and, therefore, a part of us that that person does not accept.

The secret to resolving the infinite demand is as follows:

if you were not accepted by your mother, you need to accept yourself on your own;

if you were not recognised by your father, you need to recognise yourself on your own.

It might seem difficult, but it is not if you go back to what we said earlier, and that is that on a spiritual-energetic level, the energy strategy works like the infinite demand; it is the method used to induce others to respond to the infinite demand (for example, accept me because I am a victim; recognise me because I am taciturn), and so on.

Therefore ceasing to use the energy strategy and taking energy from the Universe means resolving the infinite demand at all levels. By so doing, not only do you have energy, but you also have everything it represents, at all levels, always, without needing it from others or from outside you.

Taking energy from the Universe means always having the awareness of being what you are and the security and balance that derives from that.

11. Returning to others what belongs to them.

This paragraph deals with an issue that is of fundamental importance in a path of personal evolution. Following studies conducted in relation to the research on the Law of Delta, at a certain point it became clear that there is something within every human being that does not belong to them. Something that is not really theirs, that they borrowed from the outside and that therefore belongs to others. Depending on whether you look at it from a biological point of view or through the spiritual "pragmatism" of shamans for example, we arrive at clearly and sharply defining as *alien mind* something that has been found in all types of research and cultures that deal with human evolution, throughout all historical periods of time. Whether we call them "biological memories", "teachings absorbed from the outside", "spirits of trauma", "alien implants" or anything else is of little importance; it definitely regards something that is extraneous to the person involved.

Different methods of study, different ways of approaching the matter, different periods of time in which they developed, all seem to lead to the same result.

There is no doubt to the fact that human beings are conditioned by something outside themselves, which they need to free themselves of in order to become fully aware of their life and of reality.

There is something alien in all of us, something that belongs to others, that we do not need, whether we call it by the name of inherited biological memories or external teachings, or whether we define it using anthropomorphic descriptions. It does not matter; what is certain is that it is something that no longer makes sense to exist within us. Freeing ourselves from what we refer to herein as the extraneous mind and which can be identified with the base dissonance is one of the main elements needed to lay the foundations for taking our quantum leap.

We therefore need to be ready both to see reality from this complex point of view, and to accept to change it now and forever. Having brought the automatic responses to a conscious level makes us aware of how extraneous these responses are to us and the incongruity they create in our life.

With the help of the Law of Delta, it is easy to free ourselves of this "extraneous mind", because all we need to do is project into the brain the image which contains our own responses to each situation, definitively excluding the extraneous responses, from wherever they originate. The three brains are thus made to vibrate at the same vibrational intensity.

The influence of the "extraneous mind", in fact, leads the cerebral cortex to having -and therefore to transmitting- different information from that contained in the midbrain and brainstem. What makes us certain that the discrepancy, therefore the "external" influence is to be searched for in the area of the cerebral cortex derives from the fact that the latter contains all the truths we have acquired, and these are often in contrast with our biological needs. What we need to do therefore is change the limiting programs we have learned and that are, by definition, outside of ourselves.

From this time on and from this more complex point of view, the sense of what we have seen above becomes clear and, in particular:

The infinite demand

Our way of taking energy

Our base fear

Our limiting beliefs and convictions

In fact, identifying these points as common to everyone is proof that there are equally common mechanisms through which everyone moves and which not only do we not master, but of which most of the time we are not even aware. They repeat themselves time after time, like a nursery rhyme that was learnt and repeated endlessly by those who came before us and by those around us, with just the tiniest variation from person to person, but that in essence always remains the same for everyone.

A nursery rhyme learnt and memorised: that is what the lives of human beings are often reduced to.

The same way of living for everyone in every part of the world, with the same desires, the same aspirations, the same problems, the same malaise, the same angst...

The same sadness in everybody's eyes...

The best solution to this constant repetition of things, situations, people, models, habits, dissonances, fears, malaise – in a single word disharmonies – is to free ourselves from everything that does not belong to us and start to live our own life, different, beautiful, joyous, happy...magic. The solution is to free from what does not belong to us – not anymore – and abandoning it forever.

To this end, it is wise to clarify something that at first glance may appear to be unpleasant but that, once fully understood, will appear beneficial to the reader's biological brain. The day in which we free ourselves of our dissonance forever will initially seem to be a very sad day. Because on that day, we will be forced to rely only on our own strength which, after so many thousands of years of repeating the same nursery rhyme, will be practically null. Nothing in fact will be left to tell us what it is right to do or what it is not right to do, how to behave, how to live our life. After such a long time spent thinking like other people do, and satisfying needs extraneous to ours, and desiring other people's wishes…the freedom to be able to satisfy our own personal needs, to desire, but especially the ease of obtaining, will shock us, because we have no idea as to how to manage all of this, nor is there anyone who can teach it to us.

However, that is what we must do if we want to start living our life. We must all give ourselves permission to be the sole point of reference for ourselves. This means freeing ourselves forever from any form of judgment, both of ourselves and of others and, as a consequence, since this applies to all the sectors of our life, what will happen will also mean freeing ourselves forever from people who judge, because when we change, we will cease attracting them into our lives. By taking this step and re-appropriating ourselves of our minds, we are implicitly accepting the fact that all aspects of our personal life, whatever these may be, are perfect exactly the way they are.

And they really are, because a result can only be subject to judgment when compared to a reference outside of ourselves.

12. The Doubt of the Extraneous Mind.

This paragraph will illustrate the practical way of freeing ourselves forever from the domination of external dissonance by identifying the main facet through which it works within us and which is herein defined, tout court, as the extraneous mind.

If we use an anthropomorphic image of dissonance and call it the "extraneous mind", we have to imagine that (in order to keep a person under its domination, particularly when the person shows signs of wanting to free themselves of it), this energy behaves in such a way as to pivot on a particular aspect of their life – which varies from person to person – insinuating doubt precisely in that aspect and leading the person to abandon the path of evolution and change that they had set out on. It is extremely simple to identify which is the most vulnerable field of a person's life, on which it is easy to pivot in order to insinuate doubt. Doubt in fact always creeps in through the fissure that arises between what lies at the basis of the person's life goal, (taken as the reason for which the person is actually alive at this particular moment in time, in other words their sense-project), and what the person is afraid could happen to them if they dedicated themselves entirely to their sense-project. Doubt behaves like a recurring phobia, and to a person actually represents the most frightening fear of all.

Knowing our "weak spot" makes it possible to enter into the fear that it entails, deliberately. We personally choose to make happen what we fear might happen to us as a result of external factors and like this it ceases to be frightening or terrible to the brain. It will be a symbolic act, but the brain will register it like a sequence that really did take place and which we survived, and from that time on the fear and the doubt will be abolished forever.

Entering into fear by performing symbolic acts has a very strong effect on the human brain, and it always works, immediately improving people's lives. However, results are only achieved on some of the more external levels, such as the corporeal and material ones. If, as generally happens, the "doubt" lies at a more profound level, at the level that can be called spiritual, it will re-present itself after a certain period of time, throwing the person back into dissonance.

13. The Extraneous Mind and its Insignificant Motives.

The Law of Delta makes it possible to immediately resolve the onset of doubt given by the "extraneous mind", i.e. from the dissonance triggered by one of the three base fears of mankind. Through this Law, the brain places itself in the condition of considering the motives that led to the onset of the dissonance as being insignificant. It is important to understand what we really mean by "insignificant". In its simplicity, in fact, this word carries with it the great principle by which everything that takes place in the Universe is perfect in and of itself. This refers to the ability of serenely facing any occurrence that falls outside of human expectations.

It is the art of facing the Universe without wavering or being overcome by its manifestations. Not being strong and hard, but instead full of reverential awe; giving oneself to the Universe, and performing one's task within its Great Plan. From this perspective, it becomes clear that the phrase "insignificant grounds" does not mean taking the importance away from a bill to be paid or a tax collection notice in which unjust demands have been made, nor does it mean simply ignoring them; what it does mean is proffering deep trust in Universe, knowing it will ensure you have exactly what you need: the money or the ability to explain a mistake, or anything else required in order to resolve it. This is because personal tranquillity is important in order for each of us to perform the task we were given in the world and during this historical period of time: Evolve.

Therefore, the sole occupation for human beings should be that of performing their life goal, because everything else arrives with it. The Universe provides whatever is needed in order to sort all things out, so that we can all fulfil our life goal to an increasingly greater and better extent.

That is why all the other reasons which fall outside of this, are simply insignificant.

By deeply understanding this concept, we find ourselves – in the space of a very few days – putting all things in our existence in their right place, without overcharging situations which in and of themselves are only marginally important in the lives of evolving beings with expectations, resentment or any other energy.

CHAPTER V

MAKING THE BEST USE OF THE OLD QUANTUM REALITY

After having identified it, we will need to learn how to make the best use of the quantum reality we find ourselves in, because this is the one we have lived in for the most time during our life.

In fact, we need to experience, accept and recognise our reality to be sure we want to change it. We need to learn to live in the world, using everything it offers and making the best use of all the potential inherent in our life. We must once again learn the original way of living our life. This knowledge is innate in all of us, but most of us have unlearnt it, precisely as a result of dissonance, fear, behaviour, convictions.

Therefore, in order to live the quantum reality within which we have moved during the entire course of our life, we must abandon everything that has been reason for malaise and activate parts of our brain that have been inactive for thousands of years. It is essential that we throw ourselves headlong into the potential of the Infinite, learning how to take our energy from that never-ending source which is the Universe. The time has now come, for human beings, to begin desiring and creating once again.

Only by experiencing all of this will we be able to tell ourselves if and what we want to change in our lives.

Taking Energy
1. from the Universe.

Seeing the beauty of the essence of all things is how to draw from the infinite Energy of the Universe.

This means learning to see the intrinsic beauty in what we have before us, whether they are things, situations or living beings. Their beauty is what keeps them alive.

Observe that thing or person or situation with the intention of seeing its sense, because that is what makes it exist. Succeeding in seeing the intrinsic beauty in things can require years of meditation and philosophical or spiritual research, remote knowledge, in-depth analyses, and so on.

But since one of the fundamentals of delta is that everything is One and the Universe is holographic, this paragraph will show you a way of seeing at a glance the beauty and the profound significance of what we have before us. Acquiring this knowledge enables us to interact immediately with the All, reaching the same results as years of meditation and research, and instantaneously charging us with energy.

It involves being capable of observing the fields of electromagnetic energy emitted by the vibration of a person, thing, animal or even situation. The sense of the observed subject's continuing to live on the Third dimensional Plane of Existence resides in that vibration.

We can start by observing trees and plants, because they are sufficiently stationery and can be used by anyone who is starting to vibrate at a higher level. This method can also be used with inanimate objects, because they also vibrate and emit fields of electromagnetic energy.

The steps to be schematically followed are as follows:

Place yourself in front of a tree or group of trees or, even better, if you see a wood in the distance.

Observe the outline around the shape of the trees against the sky or the rocks, wherever they stand out, squinting your eyes slightly and blurring the image.

An outline made of a whiter light will appear immediately, that follows the shape of the plants;

Concentrate your attention on that light, then deeply breathe in and out three times.

The fourth time, breathe in deeply and close your eyes before you breathe out, taking the image into yourself.

As you breathe out, keep your eyes closed and observe the image with the eyes of your mind in all of its detail: the bright green of the leaves, the majesty of the trunks, the grandeur of its height, the dew that slides silently from the branches...

Feel the feeling that the beauty you just say has on your heart and, with one last deep breath, open your eyes again.

At the end of this simple exercise, you will be charged with energy and will feel totally independent. Ready to give emotions to others without fear, because you feel free from the risk of losing something for having given it away. You will gradually realise that energy self-regenerates, because the act itself of giving it triggers a mechanism of constant exchange that is learnt immediately by the brain, resulting in even more energy.

When the vibrations have finally been raised and kept high, it will no longer be necessary to recharge yourself, because you will always be in connection with the energy of the Universe.

2. Learning to desire.

Energy - Sense - Form.

Without these three elements, nothing exists. Each time something is created in the third Plane of Existence, in order for it to exist and persist, there must be a sense to its existence and a creative energy at the root of it, in other words an ability to conceive of it, both at the level of pure imagination and at the level of perception. The initial threesome can therefore be re-written as follows:

Need - Sense-project - Object.

This is the fundamental sequential structure for material creation to occur. Seen under this light, the mechanism appears extremely simple: a human being conceives of and perceives a need; he then acts to respond to it finding the sense of what must be created in order to satisfy that need, after which he realises it or, more precisely, he creates something that serves to fulfil the need. This means that the sense of the object created by means of the project is that of satisfying a need.

The element that moves the whole mechanism and that makes something exist in life is no longer need, but desire. For thousands of years, after having conceived of a need and before moving on to the practical formulation of the sense-project, man has put in place a sort of speculative process that has led him to precisely identify the need, by means of desire. If, for example, the need is that of eating, whereas originally this was enough to begin a project that had the sense of procuring food, an intermediate phase arose subsequently in which man focused on the need and established whether it was for sweet or salty food, for meat or for bread... This new passage corresponds to what is commonly referred to as "desire". Therefore, the current sequence can now be rewritten as follows:

Desire - Sense-project - Fulfilment.

Scientific experiments conducted with the aim of studying particles of matter have shown that mere observation is capable of changing reality. It has in fact been proved that the eye of the scientist observing the particle under the microscope is capable of making it change shape. If the results of this discovery are transposed and applied to the last sequence regarding the law of creation in the third Dimensional Plane, we can state that wanting something to happen verifies its occurrence, and is therefore already the start of the occurrence itself.

To paraphrase the supreme poet Dante Alighieri, we can say that desire is what *moves the Sun and the other stars*... in fact, at the basis of the feeling of Love, is desire.

The mistake that people generally make is wanting something that is already ready, skipping the steps of creative desire and forgetting the prime need that generated the desire itself.

A practical example could be useful for better understanding this concept: if the need involved is that of getting to a place, instead of using what we call "creative desire" which, in this case, would be "I want to be in that place", human beings - particularly during this last century - use a pre-packaged desire of the sort "I want to have a car to go to that place", or "I want to have an airplane ticket to go to that place". Like this, they give the desire an already identified and therefore old and preconceived connotation, instead of being open to anything that could lead to the base desire being fulfilled. In this specific case, we can say they refer to preconceived and pre-existent ways for achieving the desire, based on external models that are not suited for the person expressing it, because he is asking for the desire to be fulfilled by means of schemes that do not belong to him.

In a similar situation, what happens most of the time is that the base desire is achieved - because the Universe always gives everyone what they desire - but the person is not satisfied with it. To the contrary, most of the time he does not even realise that the desire has been fulfilled, because he had already imagined it would be achieved according to a certain scheme, appointing it as the sole possibility to achieve fulfilment. Therefore, if the desire is satisfied in a different manner, the person's brain is not capable of recognising it. This means that their mind had closed itself within the framework of the acquired image and did not recognise any others. Not only, it may happen that since the image in the form that the person expected the desire to be fulfilled was borrowed from the outside and, therefore, belongs to others, the person does not recognise it even when it does take place in the form they did expect! This can happen because the person who expressed it did not have the image of it in their brain, or at least not at a profound level, similarly to a photograph stuck on the area of the person's cerebral cortex. They will thus say they were distracted or that they did not recognise the opportunity that had come their way, but in actual fact they were totally incapable of recognising it. Therefore, the only image that the brain of the person expressing the desire recognises is that of pure desire. In our example, the only image that the brain recognises will be "being in that place".

Pure desire is the only one that can be satisfied, and in order to express it we must free ourselves of conditionings, beliefs and images that come from the outside. It is important to emphasize once again that in order to express pure desire, our own personal real and structural needs must be respected. By doing so, we will also be the creators of a great act of love that shows its respect for all the possibilities that the Universe offers, even those that are still "unknown".

To undertake this act of love, in other words to free ourselves of the conditionings, beliefs and convictions which are unsuited for fulfilling our needs and desires, we must first of all free ourselves of the automatic responses present in our brain that are no longer of use to us in this life - as we saw in the previous paragraphs - and then learn to formulate desire in such a way that our own brain can recognise it.

3. Coherently formulating all our desires.

The moment has thus come to put all the individual things we have discussed until now together.

We have learned how to desire: by expressing the creative desire that arises from pure need. We must now re-learn how what we desire can be obtained.

To do this, take a dream, formulated as a pure desire - not an induced one or, worse still, a deduced one - and, closing your eyes, think intensely of what you desire until you feel like the dream itself. This passage is extremely important, because if for example you wanted to take a trip, you must succeed in feeling like the trip: not as though you were taking the trip, but as though you were the trip itself. Understandably, it is not easy at first, but with a bit of concentration and discipline anyone can manage.

At the same time, as you practice being the desire itself, you should:

1. Avoid thinking, even if just for an instant, of anything contrary to this; for example do not think "I will never manage to take a trip" or anything of the sort;

2. Often think of yourself doing things that are typical of a traveller, such as "I will take that suitcase because I want to take these clothes..."

3. Feel the joy you feel when you arrive at the destination of the trip. Imagine yourself as you discover the city, the museums, as you swim in the crystalline waters of the place, etc.

4. Moreover, each day, in your daily life, take care to do everything that can be done during that day. This means properly filling the position that you occupy in your life and being efficient in every single thing you do. For example, do not distract yourself from your office work in order to concentrate on the trip that you want to take and, even more importantly, do not think that the job you are doing is unpleasant and that you could be on holiday. The best thought would be "I am here now and I am happy to be doing my job; then, when I finish, I will dedicate my time to programming my trip to the best of my ability. . .".

When you succeed in feeling this, in that precise instant, the desire begins to manifest and situations naturally line themselves up to support you in fulfilling your desire. All this can take place because you have entered into sufficient contact with the Universe. This is an empirical method and, as such, timings for fulfilment will be those connected to human linear time. More time will therefore be needed before it happens, but happen it will, because this method also works.

By knowing and using the Law of Delta, it is much easier for pure desire to be fulfilled, because all you must do is correctly formulate the desire by transposing it on a Dimensional Plane of Existence other than the third, to then immediately obtain what you desire on your own physical plane of reality.

From the instant you formulate the desire everything in your life begins to fall into place so it can be fulfilled. It is important to emphasize that in this case delta works for the immediate realisation of the event, whereas with the empirical method, this preparation requires more time.

4. Freeing the cerebellum for the conscious use of delta waves.

Once its structure has been identified and freed from everything that can, in a word, be called the "alien mind", everyone can learn the delta method for projecting images and creating reality. In order to do this, the brain must first be prepared which, for centuries, has lost the habit of consciously using some of its parts, such as the Silence Zone and the Cerebellum. The Cerebellum can be imagined as though obscured by the "dust of centuries", as if it were covered by a very light and extremely fine veil that prevented its full and perfect use, in all of its innermost potential. It can be thought of like the dust on the gears of the most delicate watch: the watch continues to work, but not in a perfectly synchronised manner.

In order to render these parts of the human brain fully functional and capable of developing all their capacities, it is important to remove the "dust of centuries" and make this Zone become perfectly active. The way to do this is peculiar yet simple at the same time and, once the cerebellum has been cleaned, makes it possible to access the Psychic Power where the creation of physical-material reality takes place.

5. The Centre of Psychic Power.

At this point, brief mention must be made regarding the points of the brain on which we must work to release the ability to consciously use the delta brain waves.

It is common knowledge that the left hemisphere of the human brain is, amongst other things, home to the characteristic of activity and electricity.

Despite this scientific definition and delimitation into zones, however, already at this point of the awareness path and more still as we proceed with what can be referred to as the deep-cleaning of the being, we begin to work with the whole brain at the same time, and each specific function point can be identified in the exact point. In the case of delta waves, the specific function lies in the front left-hand lobe.

By accessing this part of the brain and irradiating it with light, the passageway is opened to the memories stored in that zone and that up until then had not been available in a conscious manner. These memories partially concern past times "elsewhere", (i.e. between one bodily life and the other, when the soul was still not incarnated), and partially the ability of the man-God to create his own reality immediately, while he sees it in the image that is consciously projected precisely in that part of the brain.

Remember that, as I said previously, delta waves belong to all Planes of Existence. That is why those who use them are able to immediately create their reality. By illuminating the Centre of Psychic Power where the delta waves reside, we are then able to restore all the Laws of the Universe to our conscious memory (because they have always been present in our unconscious), to then use them to our benefit and to that of others.

6. The Great Abandon.

Once we are free from dissonance and fear, the time for consolidating the results achieved and for obtaining greater freedom from the "mind of others" arrives. This helps us to become increasingly independent and free from any conditioning. We feel the need to go back to finally being ourselves, freeing ourselves of all residual fears, in other words freeing ourselves forever from the base dissonance. We must therefore do what can be called the "great abandon", which consists of definitively abandoning fear, in whatever form it has presented itself up until now.

The abandon takes place by raising the vibrations of the Being to the highest level. The longer we succeed in maintaining the vibrations of all of our being high, the faster we can free ourselves definitively of the dissonance. This requires the ability to hook-up for a few minutes to the cohesion energy of our energy body, in other words to our Personal Emanation. This energy can be identified in the chemical and nuclear bonds that hold together the molecules that make up the human body. These in fact are the bonds that assemble the molecules giving them their well-known visible shape, which has always been referred to anthropomorphic. The energy of the chemical and nuclear bonds is what holds reality together, revealing it as an agglomerate of matter.

It is a question, therefore, of hooking-up to our own energy of cohesion, which is the maximum force currently present in the third dimensional Plane of Existence. By consciously using delta waves, we can hook-up to the maximum energy, raising our vibrations to the highest level known till now - which is that of light - and protract it for a long enough time to conclude the great abandon.

CHAPTER VI

PREPARING FOR THE NEW QUANTUM REALITY

Once we have freed ourselves of everything that belonged to others, it becomes easy to identify the new quantum reality that we really want to live.

In this chapter, we will talk about some of the ways that enable the new individual to create his own reality, becoming independent and free from the needs of others. It becomes possible to create the reality that best fulfils his "biological" needs, free from acquired, inherited and learned constructions and archetypes...

The new individual can go back to being in balance with the Universe and to using all the tools It has placed at the disposal of human beings, for their utmost good. He can learn of everything that is contained not only in his personal vibration, but also in the universal vibration. He can learn of the universal Whole. Once he has returned all the ancient knowledge to himself and begun to use it once again, he will be ready to build and live his own new quantum reality.

1. Being One with the Universe.

Having vibrated at the maximum vibration, at least for a few minutes, as well as having forever freed the person from all residual fears, has given him the image of maximum unitary vibration. In fact, the vibration took place both within his energetic and within his physical body, and in it the two bodies merged becoming a sole one. The energy hook-up therefore gave the brain the image of oneness. The presence of this image or, more precisely, the memory of it, now becomes very important, and facilitates the path of evolution.

We also said that once the process of removing dissonances, beliefs, convictions, biological memories or learnt memories has been completed, the three brains gradually line up with each other.

This means that the information contained in the cerebral cortex, in the midbrain and in the brainstem begins assuming the characteristic of oneness, so that they all carry the same vibration. This signifies that all the parts of the brain have the same information, which corresponds to the person's biological needs, and that the dissonance is over.

However, despite having acquired this oneness, people at this point still have a conception of themselves in which they are divided into levels, into parts: they feel as though they have a biological brain, a rational brain, a body, a soul, a genetic level, a historic level, a conscious, a subconscious, a higher Self... They feel as though they are fragmented into many parts and coordinating them all involves an enormous expenditure of energy.

That means that the time has come to lend total oneness to the Being, who has to go back to being One with the Universe and vibrate in unison with it, in the maximum vibration of infinite harmony. All the fragmentation into parts of which the Being had been subject to, despite it having been extremely useful up until then, must go back to transforming itself into oneness with himself and therefore with the All of which he is part. After having recovered unity with the Universe, everything changes in the human being: the centre of the All, the heart of his being, from now on becomes the point situated in the middle of his breastbone and distinctions between mind, body, soul, inner person or anything else are of no further use.

By consciously using delta waves, the image of One with the Universe can be restored to the person's brain. After having reacquired this image, the person immediately feels in axis with the Universe, in constant and continuous tune with himself as he is and with the Universe, and the Universe with him.

The new way of feeling is a prelude to great change, which leads to a different and total immediate awareness of the All. After thousands of years of blurriness and disharmony, everything now comes clearly to awareness and we enter into the harmony of the All.

2. Releasing the Being's heart.

Since we said that when balance with the All has been restored, the point of contact between the human being and the Universe is concentrated in the central part of his breastbone, in the area which some disciplines refer to as the "Fourth Chakra", it is necessary to ensure that this part is particularly cleansed of the crystallisation of any past emotions that may have remained. The whole central zone of the sternum must therefore be cleaned. If, during meditation, we concentrate on what can be called the Being's focal point before it has been completely cleaned, it is easy to find ourselves seeing images that flow before the mind, as though they were frames from previous lives.

This book does not intend to sustain or contrast the theory of reincarnation because, whether it is a question of lives lived previously by the same person who is observing them, or lives that belonged to those who passed down their biological memories to him, one thing is for sure and that is that if considered from the point of view of linear time, they are in any case previous lives. They are fragments of the past -that person's or somebody else's-, still crystallised in that point of the Being, that should be removed using the fastest means possible.

We cannot conceive of removing them one by one, because stories often run before the mind one after the other causing a great sense of fatigue and, at times, even of profound suffering. The non-concluded emotional sequences are always present in the biological brain, which does not conceive of linear time, and it thus relives them with the same intensity with which it experienced them while they were happening. To the brain, everything happens at the present time, therefore it is of fundamental importance to definitively remove all remaining crystallisations. Removing the crystallisations of still-open emotional sequences, in fact, is the equivalent of definitively concluding these sequences and in the best way possible. As we said above, there is no linear time for the human brain, therefore concluding the emotional sequences not only benefits the person who frees themselves of them, but also those from whom he inherited them as non-concluded, and those to whom he will now leave them concluded in heritage. This is the arcane yet not fully understood reason that led the Ancient Peoples to calling the point situated in the centre of the breastbone the *"Centre of the Creator Vital Breath"*. In it, in fact, lies the possibility of creating new Beings, free and in balance, in the harmony of the emotional exchanges that constitute life itself.

By consciously using delta waves, all the crystallisations can be dissolved, concluding all the emotional sequences all at once. A real change in the person's DNA thus takes place, which will also be left in legacy to his descendants. This change is the reason for which, once freed of all crystallisations, we have the profound perception of having rediscovered our Body of Light. From this moment on, the brain begins to perceive the entire Being as if made entirely of white-golden light. The image of the body that irradiates a ray of golden light from the centre of its chest, which expands radially the further it moves away from it, can now be found in the brain. We become aware that this ray of light is be the way to communicate with all dimensional Planes, with no exception.

The Body of Light is now ready to begin its new path, and the steps taken till now were necessary to reawaken it.

3. Filling the Void of the Extraneous Mind.

When we are witness to the change in our DNA and to the final passage to what, in time and in various disciplines, has been called the "Body of Light", we feel the impelling need to fill the void that is left when the Extraneous Mind, tout court, is abandoned.

For such a long time, it had been the Master of our reality, and had thus given us so many worries and preoccupations, entirely occupying our lives that, paradoxically, when it is finally and totally abandoned, we miss it. We are de-structured and do not know how to manage our new reality and the unknown life that we sense is rapidly approaching. It is therefore fundamental to fill the void left by the previous mind with our own images, ones that are suited for satisfying our real needs and ensuring the pleasantness of our life. These images were partly reawakened in the brain by changes in use of the automatic responses, but their range must be expanded in order to fill the substantial void left in the cells by the lack of the chemical exchanges that were activated by our fears. The emotional range can be expanded by restoring to the brain all the images -ousted as a result of dissonances-, but fortunately still present in our Personal Vibration and that, in fact, can be grossly compared to the recycle bin present on the desktop of a computer.

Imagine having a computer with a hard disk that has used up all its memory as a result of particularly heavy files. In order to free space so you can work with the computer, you take the less important files and throw them into the recycle bin on your desktop. Similarly, as a result of inherited memories and information from the outside, but especially as a result of the fears that saturate cells, the memory of the brain became so full that we freed space by shifting the images we felt were less important for our lives to our own Vibration. However, since we were immersed in our dissonances when we did this, we moved the very images that satisfied our needs into the "Recycle bin", keeping the images of the dissonance instead, in other words those regarding the "Extraneous Mind".

The beauty of this mechanism lies precisely in our Personal Vibration, which continues to behave exactly like the recycle bin on the desktop, thus making it possible, once the memory of the hard disk has been reset of all dissonance, to recuperate and reinstall the images that had been thrown into it. Once the images have been recovered, the disorientation created by the loss of all the information believed to apply until then ceases, and the approach to our new quantum reality begins.

4. Learning our Personal Vibration.

Delta enables us to create our "new rational mind", with our new perception.

We therefore need to understand and find our Personal Vibration, with the aim of "filling" the rational mind left empty and disoriented when the "Extraneous Mind" or, if we prefer, the automatic responses, were abandoned.

By consciously using delta waves, we can visualise our Personal Vibration. After appropriate preparation, and by means of a particular deep meditation, everyone can access the Centre of their Psychic Power. Everyone who has been prepared for that meditation will be capable of consciously using and handling the waves produced there. It will subsequently suffice simply to think of their Personal Vibration and to want to see the absolutely best image for themselves in a certain sector of their lives.

A space-time cross-point will light up in it, and you will be able to see the image contained therein. All you need to do is look at it and memorise it, without saying a word: everything takes place through the images, all you need to do is watch.

There is nothing more beautiful and fascinating than seeing our own Emanation for the first time. It appears in the form we described, and the lines are extremely thin, infinitesimal, hardly perceptible and extremely luminous. The space-time cross-points are also points of light which, on command, open like luminous tunnels, making it possible to cross over the space and the time that seemingly separate you from the image you asked to see. It will be like undertaking a voyage between the infinite galaxies of the Universe, whereas in actual fact we are travelling within ourselves and within all our quantum possibilities, choosing each time to enter one of them. If, in fact, we imagined joining all the points that originated from the space-time cross-points, we would obtain nothing less that the photon trails of all the quantum possibilities of the Being to whom the Emanation belongs...

Love and Beauty merge in the contemplation of the image of our Emanation, and it is then that we are able to comprehend the paradigm by which Beauty is the bridge between man and the divine, it is the way paved between Sky and Earth, the way between the Sky and the Earth; it is the Form of Love...

From this time on, the brain will begin to search for and create Beauty, always and everywhere.

Learning the Universal Vibration.

In the same way as above, we can also come to understand the Universal Vibration, and access the images contained inside the dimensional cross-points.

The utility of this comprehension is different, because it aims at expanding the awareness of human beings. Access to the Universal Vibration brings images to the brain that have been lost to the human mind for thousands of years, to such an extent that trace of them has even been cancelled from their Personal Vibration. The loss of these images can always be traced to the convention of space and time to which the human brain has chosen to adhere over the past few millennia. Because images are lost each time dissonance arises, (in other words when the three parts of the brain itself lose their alignment and therefore when biological needs and learnt truths also become unaligned) we can deduce the extreme state of disharmony between the conventions of space and time and the biological brain. Moreover, it also implies that these conventions belong solely to the sphere of the cerebral cortex, in other words to the memories learnt.

From all this we can deduce that human beings were originally born to live in circular time, and only afterwards did they have to adapt to life in linear time, with great effort and expenditure of energy.

We can even affirm that the great initial dissonance is one and the same for all human beings: life in linear time. Only by placing this great, single dissonance at the basis of all disharmonies can we understand all of humanity's fears, which have the death of the body as its sole terrible point of arrival. The end of life on one dimensional Plane of Existence. These fears would not exist without death, the direct consequence of the conception of linear time.

CHAPTER VII

CHANGING CELLULAR INFORMATION

Within a path of evolution and awareness such as the one described in this book, harmony and balance must be reached at all levels, with no exception. In order to reach a state of harmonious balance with the Universe, it is absolutely necessary to make sure that the person's Existence is not jeopardised on any dimensional Plane. To better understand this concept, imagine for example all the various disciplines which over the centuries have encouraged sacrificing the bodily aspect in favour of the spiritual one. This situation does not seem very distant in time, in that even in every-day life, the bodily aspect continues to be discredited in favour of the intellectual one, considered in some way to be of higher value. In this historical moment in time, as we are abandoning the old grid and moving towards the new one, it is important to have harmony of the all, and therefore to give the right balance to the bodily existence of human beings. Human beings exist on the third Dimensional Place, thanks to the aggregation of particles and of matter. That is all there is to it. It is a reality that cannot be argued with, but simply accepted as part of what the Universe has made. Human beings exist on this dimensional plane also on a bodily level. This level is obviously important, and so we need to look after it, preserve it and prepare it for the passage, as we must do for the other levels. For human beings, the passage to multidimensionality also includes the body.

This is the great novelty.

Many Beings exist in the Universe on various Planes of Existence, but only human beings exist on a bodily level. This is the reason for which the Beings who must help the planet in this passage are alive at this time and incarnated in a body. In fact, only those who are equipped with a body can fully comprehend what the various steps are that must be taken and the evolution that needs to be undertaken by themselves and by others so that the passage of the Beings to which access to multidimensionality is allowed together with their bodies – human beings precisely – can take place in the best way possible.

Looking at it from this new perspective the human body, which has always been considered an enormous hindrance to evolution, viewed only in its highest expression solely and exclusively of an intellectual, emotional and spiritual nature, now –to the contrary- becomes an element of prime importance.

It is so relevant as to render the active intervention of other Beings in the Universe impossible. The latter can in fact perform a role in transmitting other knowledge of help in reawakening human potential. They cannot however participate actively, both because they do not have a physical body that enables them to feel the correspondence on a bodily level to the external strain of change, and because they must respect the principle of free will that permeates every single thing in the Universe; that is why human evolution is entrusted to human beings and it will be their exclusive task.

Given its fundamental importance, the essential parts regarding the theme of the protection and evolution of the physical body will therefore be discussed in this chapter. From this study, it will be easy to deduce the direction towards which the dimensional plane, in which the body itself exists, is moving.

1. Being/s in flight.

Those who begin to create the new reality of their life in all sectors, with no exception, find themselves "in flight", because they have taken the leap to make the quantum jump. All Beings in flight will complete the jump, because everything that constitutes the past in their life is now over forever. Those who start to "expand their awareness" temporarily find themselves dazed in relation to the reality that surrounds them, because they are moving away from it to create the new reality. The new reality is not within their reach yet, when they realise that the old reality does not exist anymore. For a while, they will need to adapt, but then everyone will find the way to interact with the surrounding reality on their own. Then and only then can the path to fulfilling one's own life goal be undertaken. That will be when we realise that, in the past, they have always changed their reality on the basis of the current one. In other words, they have always made requests for change which, from their point of view, represented an improvement with respect to what already existed.

When, however, we prepare ourselves to change our reality in an absolute sense, without terms of comparison with the past, we must be ready to live without any schemes, understand that all things can be used to our advantage, satisfy our needs, whether or not they are part of our original structure. When we take a decision of this kind, and we begin to put it into practice, everything that belongs to the old grid changes, because a decision of this kind implies the final passage to the new reality. The current historic conjunction is such that taking the quantum leap in this moment in time not only signifies choosing the photonic trail on which we want to live, but also the automatic passage to the new grid and therefore to multidimensionality. If there were a scale of values capable of assessing the power of a quantum leap, we could without doubt say that a quantum leap made in the historical period of time in which we are in now corresponds to the highest leap possible. This leap is in fact capable of taking the person who chooses to make it not only onto another quantum line of reality of their life, but actually onto another quantum line of terrestrial, and therefore universal, reality.

Everything is changing, everything is in transition already in and of itself; therefore even the choice to change and the consequent achievement of this change by every single person substantially influences the acceleration of the change of the All.

The time has come for all human beings to learn to be "jugglers", to be Masters of Life, to move within it in a constantly new and fun way, to create infinite beauty and harmony... This is the moment for absolute freedom, for the total abandon of all limits, for being in tune with the All, for being in the Universal synchrony, in circular time, in the Infinite... it is the starting time for the new path...

The quantum leap in this historical moment of time is: existing on infinite levels in infinite places, and Infinitely. Now, those who have followed this path of awareness are capable of understanding it.

2. Modulating the concept of Infinite.

When we reach such a deep extent of knowledge, we need to free ourselves of all limits, whatever these may be.

When you are about to set out on your journey into the new quantum possibility you have chosen for your life, new concepts need to be introduced that go beyond past and present human knowledge, that are only known to human beings in their abstract form. They have never been fully assimilated by their Being, so much so that no image of them can be found, either in the Personal Emanations of people, or in the Emanation of the Earth. New images must therefore by projected into the brain and human awareness expanded. These concepts must therefore be taken from the Universal Vibration. That is because the human brain has been programmed to recognise and read images that, in some way, it has knowledge of, either directly or indirectly. In other words, the brain is capable of reading images that it itself is used to seeing, or that someone else read before it or, even, that someone else is reading in the instant in which they find themselves next to them. It is easy to understand this method of operation, especially after having discussed biological memories and the laws of cellular communication. In fact, we have said several times that "cell speaks to cell". This kind of operation explains how the human brain works.

To better understand this concept, let us take the example of something that happened to a six-year old girl.

As it goes, this girl was born in the mountains and had always lived there. Because of this, she was familiar with the mountains from the day she was born, but had to wait till the age of six to see the sea for the first time, when she went on holiday to her aunt's and her uncle's accompanied by her elder sister. The girl observed the bright and vivid colours and memorised all the flavours and images of that first day of her holiday. Finally, after a wonderful and, in her opinion, extremely long trip, they arrived at the pinewoods along the beach and, while the grown-ups were busy unpacking and preparing the house, the older sister picked her little sister up and from the pinewoods showed her the sea in the distance. She pointed at it with her finger and eagerly said "can you see it?" She was bewildered to hear the little one's answer, because she said "no". And in fact she couldn't see it, she just saw the sky. The older sister kept insisting and the little one saw that the sky became darker and darker as it descended to touch the horizon. She explained to her older sister that all she saw was this difference, and the older sister explained to her that that darker part actually was the sea.

From then on, to the child the sea became "the darkest strip of the sky before it meets the earth".

This means that the child was born without the image of the sea, but that that image was in any case present in her Personal Vibration, because someone close to her did have it. Her sister in fact had already learnt it, and therefore managed to recall it in her, making it pass from her own Emanation directly to the her brain which, within a short space of time, was able to decode it and place that image in her memory. That is what happens continuously to the biological brain when it sees images: it decodes them, according to pre-existing schemes within the person or within groups of people, and then it memorises them. But if for thousands of years no-one had decoded the image of something that appears before our very eyes, then we will not be able to see it. It simply will not exist. In the case of our example, the sea would not have existed, for the girl there would only have been sky. That is why this paragraph discusses images that do not exist in the human mind and that mostly concern the different conception of time and of space, particularly the circularity of time and universal synchrony.

The linearity of time in which the Personal Vibration has been bridled for eons has caused these concepts to be ousted from the Emanation itself, and replaced with concepts that relate to "finite". The current loosening of the cohesion of the Terrestrial Vibration has made it possible for us to connect to the Universal Vibration that – as we said previously – is made up of dimensional intersections and where, in fact, neither time nor space exist, and the intersection points are made of planes and vibrations. Once we "enter" the Universal Vibration, we take the image and project it into the Personal Emanation and, consequently, into that person's brain.

All the issues discussed till now lead to the natural conclusion that, in order to expand awareness, the first image to be projected is that of "Infinite". Needless to say, this can only be done in delta. More precisely, it involves going to the Universal Vibration, to a specific dimensional point of intersection, taking the image of Infinite and projecting it into the each one's Personal Vibration.

Another reason for which the expansion of awareness begins from the concept of Infinite is that in the new common quantum reality, we must be able to conceive of Infinite Place, Infinite Space, Infinite Change, Infinite Places, Infinite Worlds, Infinite Interactions… in other words, we need to be able to conceive of Infinite life in the world.

Why should we conceive of life this way?

If this concept is lacking, (and up until now it has been for human beings), finite concepts that bear the idea that make an infinite life seem pointless, and this is because life extended to the Infinite is perceived as repetitive and therefore useless. Thanks to the new concept of "Infinite" and, especially, to the awareness of this acquired knowledge, we can now expand our horizons and start to abandon the concept of finite, which up until now has permeated human life on Earth, giving rise to duality and contrast.

Up until now, in fact, the human brain has only imagined "Infinite", in other words it has conceived of it as a thought, but it has never taken it to the midbrain and to the brainstem, making it become part of itself. By means of the new image of Infinite taken from the Universal Vibration, the human brain begins to behave like "Infinite". It abandons prejudices such as those that consider eternal life like a punishment. In which one believes that those who live for eternity are forced to see the people with whom they have lived depart for ever, and other similar prejudices. With the new image of Infinite, the brain expands its scenario, at least until it realises that when Infinite exists, it exists for everyone, and no-one can die before anyone else unless they expressly want to.

Then it understands the most important thing, and that is that Infinite is not the same thing as eternal. Eternal, in fact, is a sort of curse, something from which one cannot escape, a final condemnation, and therefore also finite. Whereas the prerogative of change is innate in the concept of Infinite.

Infinite is something that changes continuously, it is not stable, and therefore "living life to the Infinite" means living until one does not want to change. And, in this particular case, changing may also mean abandoning the body, going elsewhere, moving to another Plane of Existence, but also going elsewhere with one's body, or yet other different possibilities again.

Infinite possibilities.

3. Reactivating the Thymus Gland.

When people accept the concept of Infinite within themselves, making it their own, it means they are ready to live life to the Infinite. As far as this is concerned, if we choose to live in the quantum possibility of Infinite life, we can interact and give form to this reality by changing the cellular information contained in the body.

It therefore becomes necessary to preserve the Thymus Gland in all of its functionality.

If you look carefully at the exact position of the gland in the human body –behind the sternum, sitting against the pericardium–, you realise that this is the exact point in which, after restoring balance in relation to being "One with the Universe", we see the pulsating heart of the Being. The centre of ourselves from that moment on. The point in which we identify the *"Centre of the Vital Breath"*.

Taking steps to maintain the Thymus gland efficient means ensuring it does not atrophy, despite the circulation of sexual hormones. In other words, that it continues to produce lymphocytes, not only to defend the organism, but also and especially, to preserve it against ageing.

Going in sequence, if we observe the various steps indicated up to now in this path of personal evolution, it is easy to see that the person has been freed of everything that once weighed his Being down.

In a short space of time, beliefs, things, facts, people, thoughts, behaviour, dissonances... were removed, he became aware and present, but something still remains to be set right, precisely in that spot, in the centre of his chest. The removal of the crystallisations of the unfinished emotional sequences has taken place and influence him at the Spiritual level; now the part must also be reactivated on a material level, so that once again the two levels correspond with each other.

One must therefore reactivate the thymus gland and restore it to its full functionality. To succeed in this intent, the information in the cells must be changed. For thousands of years, in fact, this information establishes that from puberty onwards, the gland will begin to atrophy. But just like any other information in the human body, this information can be changed by means of the conscious use of delta waves.

That is what we meant when we said that the time would have come for also changing what up until now had been defined as biologically "structural" in human beings, and as such unchangeable.

What we are saying by this is that delta can also change the cellular information that for centuries has been believed structural, because this is only information and, as such, has been acquired and can therefore be changed. This statement has extremely far-reaching implications in the field of research, of whatever kind it may be. In fact, the possibility is surfacing of changing the information on ageing and of death by ageing, currently contained in the cells of the human body.

Which is the same thing as saying that immortality of the physical body is possible.

4. Regenerating the Thymus Gland.

By reactivating the Thymus gland by means of delta waves, it reinstates its functions, and the cells lose their memory of progressive ageing, because that memory is no longer transmitted by the gland in question. The memory of ageing however still remains, albeit slowed down considerably, as does the memory of death of the cells themselves at the end of their vital cycle. These cellular memories persist because, even if cleaned, it is the same thymus gland.

It therefore becomes necessary to regenerate it consciously, because only like this can the cellular memories connected with the thousand-year-old conventions of space and time, inherited by all the human beings or animals that live on this bodily dimensional plane of existence, be cancelled entirely and replaced by the laws regarding the other dimensions.

The Thymus gland will be visible on the Personal Vibration to those who have received the necessary preparation for profound meditation. Once the gland has been identified on the Emanation, waves of light will be sent to it (in meditation) in order for it to be regenerated.

What one can see in the awareness of the brainwaves is truly spectacular. You can see the gland, which initially presents itself ruby red in colour. All of a sudden it begins to whirl round and round in a clockwise direction until it gradually becomes shining white, finally presenting itself made of the brightest white light when it stops spinning. The light of which it consists spreads all around. Generally one stops to admire it, astonished. When you proceed with a regeneration of this kind, even after you reopen your eyes, the image of the gland made of light and its splendour within the body continue to be real in the person who performed the Deed, as though what they just saw continued to take place before their eyes, giving them the sensation of having achieved something truly great. It is enthralling to understand what took place in the brain: the image of the regenerated gland had just been created and this took place in such an intense manner that it was impressed upon the mind. The brain transmitted it to the body which, in its turn, began to change reality.

This is always the way in which reality changes:

it is enough to possess the vivid image of everything you want to do.

Obviously, as we explained in greater detail earlier, the drive behind this process is desire; it is true to say that we cannot create reality if we do not know exactly what we want. This means is that if our desire is not intense enough to push us to search for the most suitable image for ourselves of the thing we desire, then that thing cannot be created for us, either by ourselves or by others.

By looking for the image of the Thymus gland and of its regeneration, in an extremely simply way the delta operator created the regeneration of the cells of the body without the genetic information of ageing and death.

At this point, the power of the Law of Delta can be fully understood.

It is immense, immeasurable, as only the Universe can be.

5. Vibrating at the intensity of light.

Raising the cellular vibration in the human body and the ability to transmit information between cells until we make them reach the speed of light means ensuring that the cells continuously regenerate themselves from light itself, because they are capable of vibrating at the same intensity. The newly achieved vibration favours their rejuvenation and the consequent Infinite maintenance of that state, because the energy of light never ceases.

The continuous regeneration of cells also coincides with healing and with the constant and perennial maintenance of the state of harmony.

Raising the vibration of cells is very interesting and perfectly appropriate to our disclosure of the Law of Dimensions. Without this increase, in fact, the method that establishes the conscious use of delta waves, in fact, immediately affects the person on a Spiritual, Intellectual, Emotional and Sexual level, in other words on all the dimensional Planes of Existence except the Third One itself. In this latter plane, existence prevails on the person's bodily-material level, and in implementing changes enacted by means of delta waves, we have to come up against the convention of linear time. Therefore on this human level, we can see that delta works, but the change follows the timing – albeit accelerated – of the biological time required by the cells in the body for them all to have the same new information or, more precisely, to reproduce themselves with the new information.

During this stretch of time, which lasts approximately thirty days, the passage of the new information to all the cells can be cancelled or interrupted by chemical combinations with residues of the ancient weave; in order words, by emotions regarding issues of the past.

With cellular vibration and the ability for transmission elevated to the speed of light, the ancient weave can no longer interfere to alter things, because its vibration is too low. In this way, we can reach and stabilise the best states for the body, in relation to its power, flexibility, adaptability, strength, beauty...that render it capable of living infinite longevity.

In the introduction to this chapter, we explained that in delta evolution, the beauty and youth and in general the physical fitness of the body, which is the mode of existence in the world, is of particular importance. At this point, it is easy to make these concepts merge with the concept that asserts that Beauty is in Everything, and that if human beings are able to glimpse it, then they are able to draw from the energy of the Universe and raise their vibrations. And by connecting them, fully understand them in their essence. In fact, the concept of Beauty was previously given as a form of the divine, but now it becomes easy to understand it in terms of energy. Beauty is the form of maximum vibratory energy, and is present in the biological memories of all human beings. This means that when we have beauty, see beauty or simply sense beauty, we already are on the divine plane.

Beauty is what draws man to God, therefore, because if we are able to see beauty, we are able to grasp the maximum vibratory energy, which means we are vibrating at a very high level and are consequently capable of creating harmony and balance.

Once we reach a certain degree of evolution, beauty becomes an impulse of the soul; it is illumination that is reflected in the body.

It is awareness that materialises, making itself visible to the human eye through the beauty of the physical body. There can be no illumination, awareness, spiritual evolution, without it resulting in physical beauty, because only thus can we access the Oneness; only thus are we One with the Universe, we are a Complete Being who has overcome dualism and therefore contrast.

Because the Universe is the infinite sum of immeasurable Beauty. Only with Beauty is our being complete, in total harmony with ourselves. Once we are able to find Beauty within ourselves, and bring the image to the brain in a conscious manner, we will have complete access to the energy of the All, and our vibrations will be raised. This statement simply means that the electromagnetic energy that our body emanates will vibrate at a higher vibration and this will ensure that the personal and universal Emanation is always accessible on a conscious level. We have said more than once that if different images are brought to the brain, we can change our life. If this is true, it is also true that if we take the image of beauty of the physical body to our brain, it will start to raise the vibrations of all our being, because as a result of the holographic principle and the principle of oneness, the brain will perceive the beauty of the physical body as the beauty of the Universe.

And, in fact, so it is.

It follows that those who live in Beauty, with awareness and without dissonances or conditioning, have already started to raise their vibrations.

From what we said, it is easy to deduce that a path of evolution can begin right from this very point: raising one's cellular vibration and transmission capacity to the speed of light. This is the maximum possible degree of cellular vibration and transmission; therefore, since the human body is not used to all this, it will need at least two weeks in order to learn to handle the new state in which it finds itself from the time when its vibrations were raised, onwards.

6. Dialoguing with Energies.

When we begin to vibrate at a very high vibration, it becomes possible to converse with any of the Energies in the Universe and dialogue with any of them, learning and feeling anything we want. Having arrived at this degree of evolution, in fact, it is rare that a person continues to ask for or tries to obtain anything, because they have already changed their life and continuously obtain whatever they need, without having to ask for anything anymore. Now, their interest lies in learning how to properly manage their new reality, and begin working towards the wellbeing of others, with the utmost respect for any kind of choice and therefore without interfering in anybody's path, unless they are specifically asked to do so.

It is not necessary -in dialoguing with the Universal Energies-, to use verbs in their imperative form, because the brain has impressed the maximum vibration onto its cells and has abandoned fears and "feelings of inferiority" in relation to the beings of the Universe and of the Infinite Universes.

By rising to the maximum vibration, the human

being has now fully become a man-god and can dialogue with any form of energy, whether it is our desk, money, animals, beings from other dimensions, from other universes...

We are surprised when the body, or anything else with which we dialogue, answers by saying things we had never thought of before.

All this is possible, because vibrating at the highest vibration means Being Universal Love, and all things respond to Love. Always.

CHAPTER VIII

THE NEW QUANTUM WAVE

Up until now we have always talked about knowing ourselves, understanding our biological needs, our past, freeing ourselves from dissonance, looking within ourselves and looking out, sensing what the Great Plan of the Universe is in relation to our existence on the Third Plane of Existence...

In this chapter, we will access a new part: in fact, we will talk about how to create our new reality and how to progressively detach ourselves from all the old schemes. A new method will therefore be briefly outlined, which is precisely that of creating images by projecting in delta. This means creating a new scheme, but it is temporarily necessary to do so, because this scheme helps us comprehend that once we are free of schemes, all choices can be created. This is obviously a contradiction in terms, and therefore it is good to consider this an exercise, in which we can learn a method that is useful for creating something else. We obviously need to remember that because the ultimate goal of a personal evolution path is that of creating a person's utmost adaptability, integration, awareness and expression within the Universe -in other words, the utmost evolution of a Human Being- any scheme that was created at the start of this discourse must necessarily be removed before the end of it. Everybody must find their own way of abandoning every possible schematisation of reality.

The issues discussed until now have attempted to give an idea of what the total change inherent in the definition of "quantum leap" means. At this point of the explanation, we have now accepted the unavoidability and concatenation of the All, fully understanding the range of the path undertaken, and understanding that since everything is connected, we cannot change parts of our life, but all parts of it must change, in all of its aspects, because now more than ever - thanks to the slowing down of the terrestrial rotation and to the opening of the old grid - we are One with the Universe. Up until now we could have imagined changing our life, cultivating however the remote conviction of only doing so in part, in other words only changing the portions of reality considered to be particularly bothersome. But when we have access to knowledge, we must accept that all parts of our life change, in every tiny detail, because only thus can we take the quantum leap.

Only when the individual has understood and accepted the need for total change, can everybody proceed along the path of their conscious evolution.

1. Anchoring the new photonic trail.

Previously, we indicated an appropriate way of expressing desire and of keeping it constantly alive in our reality. It is therefore interesting to identify a way of creating our reality: expressing a series of coherently articulated desires such as to "cover" the most important aspects of our life.

The sectors singled out are generally work, sentimental relationship, family, social relationships, articulated for fulfilment according to the scheme:

Desire - correct request - reality.

Thereafter, it is important to explain to our brains what the things are that we want to achieve in each sector, so that they become clear to it. Once this has been done, it is important to go back to the emotion that, in the past, was connected to the realisation of similar desires in those specific sectors. It will be easy to recuperate the emotions cancelled by dissonance, because they are present in the Personal Vibration. Like this, the emotions go back to being consciously present in our brain, thus fully reactivating the function of the midbrain.

When -by means of the conscious use of delta waves-, the recuperated emotions are superimposed onto the images that correspond to the desires expressed, we enter the phase of anchoring ourselves to the wave of the new quantum reality we have chosen to live, from amongst the many possible ones.

During this phase, we are able to realise that almost everything in our life has changed, and that for most of the time it is exactly as we wanted it to be. However, at the same time it is clear that for certain periods -which get shorter and shorter- we feel as though everything returns to being as it was in the past. An oscillating movement that is slightly disorienting and that is due to the fact that the new reality has not as yet been entirely stabilised. This intermediate phase, however, has its utility, in that it allows us to be certain that the choice of the new quantum reality we made really is the best one for us, leaving the old reality to surface every once in a while in order for us to be able to have a basis for comparison. The human brain, which has been caged into schemes for thousands of years, still sometimes needs to question itself a little, though increasingly rarely, and this lends a certain sense of security, despite its unpleasantness. It is an obvious contradiction that which leads the person to suffer just so they can once again find something that belonged to them in the past, but that is how it is.

Once we are firmly convinced of the choices we have made, however, the time comes to anchor the wave of the new quantum possibility and to stabilise it, so that it is definitively part of our new reality.

In order to avoid discrepancies between action and realisation, in other words so as not to fall back into linear time and distance oneself from the synchrony of the Universe in which everything is here and now, it is essential to use the Law of Delta. In fact, this involves abandoning the last inconsistencies remaining between linear time and circular time. By observing what happens in the life of those who have made the quantum leap, we can see that by their own express will everything has been created and the new quantum wave anchored. However, on a level of dimensional existence, a very thin membrane remains that separates the bodily dimension from the other Planes of existence in which the wave is anchored and the new reality unfolding. The presence of this veil indicates the persistence of a nuance of a temporal nature. It is like a sheet of tracing paper that must be removed so that everything created can manifest in our daily life. To free ourselves of this subtle membrane, we must first and foremost definitively remove from our lives the weaves of the air or the Ancient Weave as we want to call it: these are the very last fragments of the old reality. These remain like splinters, until we are ready for total change. The fragments in fact have filtered the new reality, slowing it down for our utmost good. When we are ready to change even the last things remaining, all at once, is the time to remove the thin film of the weaves of the air, freeing ourselves forever of everything superfluous. Then we are ready to take the final step towards the reality we have created.

As we said, in order to create our reality or choose from amongst the various quantum possibilities, we need to move onto other Planes of Existence, in which the conventions of space and time are absent and in which saying now, immediately, has neither meaning nor significance. We therefore need to learn how to transpose what we created in other dimensions into our own reality, with immediate effect.

The search for how to do so and the direct relationship with the implications that derive from the convention of time is one of the most fascinating and enthralling parts that can be encountered along a path of evolution. It is in fact relatively easy to anchor the wave of the new quantum possibility in our lives. However, the time required for total change in a person's reality, in all of its parts, takes on average approximately one solar year in linear time. On the other hand, managing to stabilise the quantum wave means achieving realisation on a bodily-material level within an average of twenty-four, forty-six hours. This great result is facilitated by placing the image of the new quantum reality chosen for our life directly into the double helix of our DNA.

In fact, whereas the projection in delta of the new image onto three precise points of the brain anchors the new reality created, the subsequent projection in delta of the same image into the double helix of the DNA results in the almost immediate stabilisation of this reality. With this passage, the image on its own goes to position itself in the double helix, assuming its natural consistency which is that of a group of chemical elements, of the same nature as the body, since it is already of the same nature as the Universe. To put it more simply, we can say that like this the image created has already been processed as required by the human body and that it comes in the same form of chemical components that the body is familiar with, therefore it is already present in the DNA.

This is the profound meaning of certain expressions we use to say things such as "he has business in his DNA", or "that person has become part of my DNA" or, yet again, "he has music in his DNA"...

It is a question therefore is consciously doing what is already present in our biological memories, so much as to even be transmitted through common everyday expressions.

Once again we discover that everything speaks of Everything, you just need to want to see it.

2. The irresistible desire.

Up until now, we have said how reality can be created by means of individual emotions. In other words, by connecting an emotion that "worked" in the past for that particular person, in relation to the reality to be created.

However this mechanism works precisely because it can be recovered from a past memory, and now presents itself like a scheme that has already become old and is therefore not entirely pertinent to the new reality.

While we create the new reality from scratch, without any relation to past experience, we have to find the emotion that is capable of creating the new. The emotion that it is good to use is, paradoxically, the same for everyone, and that is the emotion of *"irresistible desire"*.

It applies for everyone, indistinctly. We could almost say that irresistible desire is the thing that is common to all of mankind. It is what makes children's wishes come true. It is what ensures that at least once in their life everyone has had a wish of theirs come true.

It is however what most people have lost. In order to recover it, we must go back in our memory to a time in which we achieved something in our life that we wanted immensely, but that seemed almost impossible could come true. Once we have found that image by consciously searching through our memory, we need to go back with our minds to when we decided we wanted to do the deed that is schematised in that image and, in particular, to the emotion felt in that moment. That emotion is, without doubt, "irresistible desire", which each person feels in their own individual way, once again demonstrating how everybody is different.

Once we have remembered the emotion as a sensation, the brain memorises it and associates it with the words "irresistible desire". Therefore, from that time on, whenever you want to create something in your reality, with immediate effect, you need to put yourself in that emotion and ask for the image required.

That is enough, because the ability to create that we once knew and that we lost when we were children has been found again. I have in fact been able to observe that the cancellation of this particular emotion from people's memory always takes place at an age that falls between four and ten. What usually occurs for this to happen is that a dissonance sets in immediately after we obtained whatever it was that we wanted. A dissonance that in most cases was induced by a judgment or an action by an adult, to punish precisely the child's ability to create the reality that was most fitting to him.

At this point, we could ask ourselves why this desire must be retrieved from the past if in our new life everything that belonged to the previous quantum reality has now left. The answer is very simple: without it we would not have been able to be here. The only real drive for evolution came to human beings from the irresistible desire to evolve. Each time we talk about faith, about hope, about a better future… what we are actually talking about is a wish for change, for evolution…in a single word, about irresistible desire.

And so it still is, and so it will be.

We have to reactivate the ability to feel irresistible desire, so that each of us reacquires the ability to live like a Magical Being, because that is, in power, what every human being really is.

By using irresistible desire within the complete delta technique, everything, anything that has been created in this way, comes into the life of the person who created it within the space of two days, concrete and visible for anyone to see.

Photogenesis.

3.

What we learn to make conscious use of by means of delta waves is called Photogenesis, which means the Creation of Light. This term is usually assigned to particular plants or creatures which, like fireflies, create light within themselves and then spread it outwards.

This is what everyone can always do, once they have assimilated the Law of Delta. Using light, everyone can change their reality and create a world of Light for everyone. Being able to generate Light produces important changes in the human body, that range from cellular information, implemented through the Thymus gland, to the relationship with the most ancient biological memories.

4. Opening the Star Gates.

The ability achieved to produce light through Photogenesis is, in its turn, indicative of a person's ability to access the Universal Vibration.

As we described earlier, the Universal Vibration presents itself to the human brain as a sphere made up of intersecting horizontal and vertical circles, made up in their turn of extremely fine photonic quanta that lend these "meridians" and "parallels" the characteristic of iridescent luminosity. The dimensional points of intersection hold the images of everything that is in the Universe, including images that have never been in the human brain, such as the image of Infinite, etc. By expanding awareness, human beings are able to access and especially accelerate the "passage to multidimensionality" towards which this Planet is moving. In what way is it possible to do so?

By going back to knowing what our "life goal" is and doing it. Our life goal is the reason for which we are alive in this place and at this particular time, and in general we can say in all certainty that anyone who is following the path of delta waves has a life goal that concerns helping the Earth through this the dimensional passage with joy, grace and lightness.

By using the Universal Vibration, it therefore becomes possible - amongst other things - to know what our life goal is. Here it is possible to literally overlap the two Vibrations, the Personal one and the Universal one, in order to obtain individual information, but with the general view of the Great Plan of the Universe that regards human evolution. It is in fact possible to make the Personal Vibration rotate until it coincides with the space-time point of intersection that corresponds to the image of our life goal, with the dimensional point of intersection of the Universal Vibration regarding the same life goal, but in the entirety of the Great Plan of the Universe. In this way, where the points coincide, it will be like opening a Star Gate, which will allow us to expand the awareness of our personal goal on a Universal level.

We will be able to understand what place our own goal occupies in the Great Plan of the Universe and what it -and with it the person to whom it belongs- contributes towards creating within the All in the Great Evolution of the species. Once we have found our Life Goal, all that remains is to do it.

Do, do, do: infinitely do.

SUMMARISING

We need to start by identifying our needs, sense-project, dissonance, fears, way of responding to the external environment, personal abilities to adapt, potential and characteristics.

Once we are aware of this, we will be able to clearly identify the parts of our file that are not our own, and that we need to change in order to be in balance with ourselves.

After this, by using the delta instrument, each of us can make the changes necessary, thus changing their own life.

This instrument makes it possible to change the information contained in their cells, and create change everywhere there is disharmony. It works because it arrives at the depths of the Being and because it uses subtle energy at the highest vibration. But the most important thing that the use of delta waves gives is access to our Personal Vibration, from which we can draw the absolute image of all things. In fact, very often we do not have the image of what is necessary; for example, someone who has been sick since he was born has lost the image of his being healthy, as well as the relative image.

Thus those who for too long of a time have

been used to thinking they want things that are acceptable for others, but not for themselves, or who for too long of a time have been used to desiring the wishes of others, of publicity, of friends, of neighbours, according to the social environment in which they live... these people have now lost the image of their needs, of their desires... That is why we must first find the structure of our own real needs, and clear ourselves of limiting convictions and beliefs that were acquired or inherited, from automatic responses...

Finally, we are ready to take the quantum leap.

The instrument that we have tried to describe serves, by connecting to the All, to take the image of what it means to be in harmony and then create the project for our new life. When we reach the new level, everything takes place instantaneously, because we are in constant contact with the energy of the All and everything is immediately available to those who ask.

All the Masters of the past have said *"Ask and you shall receive"*, because Everything is at everyone's disposal of: health, money, sickness, joy, happiness, grief and sadness...

It is therefore essential for us to identify what we have asked for in our lives up until now, what we have chosen to live and what we have lived.

The use of delta will allow you to master your life...

In order to do all this, years of work and of research will no longer be necessary; everything can be done in a short space of time because it is in the universal order of things that everyone has wellbeing in every field.

The Universe has much greater projects in store for human beings than let them consume themselves in problems that arise from insignificant motives. The Universe has a Great Plan in which all of us have a place, and this plan is the evolution of the human species...

Everyone has their specific role within this Plan and everyone has free will and thus everyone has infinite possibilities before them.

It is simply up to their own awareness to choose the one they believe is best for them.

Lucia Dettori

Lucia Dettori, established and sensitive architect,has devoted several years to the study and research of spiritual issues, her main interest revolving around themes of human Evolution in relation to the Laws of the Universe.

Her attention to brainwaves has led her to formulating a theory of her own, based on principles of quantum physics and mechanics, summarised in the essay entitled "Delta, the law of dimensions", 2009.

Other publications
by the author include:
Eléne 2008
The Dream City 2010
The Song of the Cards 2013
The Parchments 2015

Lucia Dettori

Contents

Lucia Dettori